Bite-Sized
Operations Management

Synthesis Lectures on Operations Research and Applications

Bite-Sized Operations Management
Mark S. Daskin
2021

Bite-Sized Operations Management

Mark S. Daskin

www.morganclaypool.com

ISBN: 9781636392318   paperback
ISBN: 9781636392325   ebook
ISBN: 9781636392332   hardcover

DOI 10.2200/S01124ED1V01Y202108ORA001

A Publication in the Morgan & Claypool Publishers series
SYNTHESIS LECTURES ON OPERATIONS RESEARCH AND APPLICATIONS

Lecture #1
Series ISSN
ISSN pending.

Bite-Sized Operations Management

Mark S. Daskin
University of Michigan

SYNTHESIS LECTURES ON OPERATIONS RESEARCH AND APPLICATIONS #1

MORGAN & CLAYPOOL PUBLISHERS

ABSTRACT

The text is an introduction to Operations Management. Three themes are woven throughout the book: **optimization** or trying to do the best we can, managing **tradeoffs** between conflicting objectives, and dealing with **uncertainty**. After a brief introduction, the text reviews the fundamentals of probability including commonly used discrete and continuous distributions and functions of a random variable. The next major section, beginning in Chapter 7, examines optimization. The key fundamentals of optimization—inputs, decision variables, objective(s), and constraints—are introduced. Optimization is applied to linear regression, basic inventory modeling, and the newsvendor problem, which incorporates uncertain demand. Linear programming is then introduced. We show that the newsvendor problem can be cast as a network flow linear programming problem. Linear programming is then applied to the problem of redistributing empty rental vehicles (e.g., bicycles) at the end of a day and the problem of assigning students to seminars. Several chapters deal with location models as examples of both simple optimization problems and integer programming problems. The next major section focuses on queueing theory including single- and multi-server queues. This section also introduces a numerical method for solving for key performance metrics for a common class of queueing problems as well as simulation modeling. Finally, the text ends with a discussion of decision theory that again integrates notions of optimization, tradeoffs, and uncertainty analysis. The text is designed for anyone with a modest mathematical background. As such, it should be readily accessible to engineering students, economics, statistics, and mathematics majors, as well as many business students.

KEYWORDS

optimization, uncertainty, tradeoff, linear programming, integer programming, queueing theory, location modeling, simulation, decision theory

To
Babette,
Tamar, and Keren,
and all my students

Contents

List of Figures

List of Tables

Preface

This book grew out of an introductory Operations Management course that I have taught several times at the University of Michigan at Ann Arbor. The course is a seven-week introduction to operations management designed to introduce students to the field and to attract students to industrial and operations engineering. The topics that I typically cover include linear and integer programming, queueing theory, inventory modeling, simulation, and, time permitting, decision theory. The course is not designed to be theoretical, nor is this text designed that way.

The book is designed with two key objectives in mind: (1) introducing students to our field and (2) attracting them to the discipline in the hopes that they will be the next generation of researchers and practitioners of operations management.

Operations Management has a broad range of applications from the location of emergency medical services to the redesign of a campus bus network to maximize customer service while maintaining social distancing during a pandemic; from the management of inventory policies at a coffee shop to the determination of the number of doctors to have on duty in an emergency room at any point in time; from scheduling of residents during their medical training to prepositioning emergency supplies in advance of a natural disaster; from the assignment of students to discussion groups to maximize some measure of within-group diversity to the determination of priority classes of people for a new vaccine during a pandemic. I cannot think of an industry that is not impacted by operations management. In fact, one of the largest industrial engineering groups is at Disney. Who would have thought that an amusement park would be a beneficiary of operations management? Yet, this group is one of the leading groups on queueing theory and keeping customers happy while they wait for the next ride.

There are many excellent introductory operations research, operations management, or management science texts. Why then do we need yet another? There are at least three key facets to this book that differentiate it from other books on the market. First, and foremost, the book is short by design. I looked at 10 such books in my office and they averaged over 925 pages; the shortest was 675 pages and the longest weighed in (yes, I use that phrase intentionally) at over 1,400 pages. The draft of this manuscript is 157 single-spaced pages. Second, while most texts in the field present a series of methodologies and their applications, this book focusses on three core components of operations management: optimization, tradeoffs, and handling uncertainty. I believe that much of operations management can be boiled down to these three components.

Third, there will be an extensive set of supplementary material online to accompany the text, including: Excel-based exercises, traditional problems, and PowerPoint presentations.

I hope that you, the reader, have as much fun with the book as I have had writing it. More importantly, I hope that you gain as much fulfillment by working in this exciting arena as I have over the course of my career to date.

Mark S. Daskin
September 2021

Acknowledgments

The book had its genesis in a new project that my editor, Ms. Susanne Filler, suggested to me. Morgan & Claypool was starting a new book series that would be shorter books and that would include many online supplements. This fit perfectly with an idea for a book that had been slowly germinating in my mind. Many thanks to Susanne for the wonderful suggestion that motivated me to write this book.

The text is dedicated to two groups of people: my students and my immediate family. Over the course of my career, I have probably taught several thousand students in the various courses I have taught. While I hope that they have learned a bit from me, I know that I have learned a huge amount from them. The entire process of teaching is one in which I have really learned material that I thought I understood but that I only really learned when I tried to teach it to others. Also, many of the questions that students have asked over the course of my career have forced me to think about issues, problems, formulations, and models in a new light. I am truly grateful to all of my students, particularly the 20 or more Ph.D. students with whom I have been privileged to work over the course of my career. Special thanks are due to Kayse Maass and Luze Xu who took the time to read a draft of the text and to comment on it.

Finally, this book is dedicated to my immediate family: my daughters Tamar and Keren and my wife Babette. I am deeply indebted to the three of you for the support you have given me and the love you have shown me over the years.

Mark S. Daskin
September 2021

CHAPTER 1

Introduction to Operations Management

1.1 WHAT IS OPERATIONS MANAGEMENT?

Operations management is the science behind the design and management of many of the systems that we encounter in our daily lives.

When you order a new pair of pants through Amazon, operations management is at the heart of the fulfillment process as well as many other decisions that Amazon must make. These include: which items to stock, the order in which items will be displayed to customers on their website, where to stock the items, and how to deliver the items from the storage locations to your home or office or local Amazon locker.

When you go through a TSA (Transportation Security Administration) checkpoint at an airport, operations management has helped TSA determine staffing levels, how many people should be used to check IDs, how many X-ray devices to deploy, and how many checkpoints to establish within the airport. Operations management also underlies decisions TSA makes about the level of scrutiny to apply to each individual passenger.

When you place a call on your cell phone, operations management is lurking behind the scenes helping your carrier determine how many cell phone towers they need in your area and where they should be.

When you contract with a snow removal company to plow your driveway, you often have to pay for a fixed number of plowings, independent of the actual weather conditions during the upcoming season. Thus, you may lose money, or feel that you paid too much per plowing, if it is a mild winter. On the other hand, additional plowings beyond the contracted number may be more costly and so you could pay more per plowing if it is a rough winter. Operations management can help you decide how many plowings to contract for. Similar decisions have to be made by OEMs (original equipment manufacturers) in contracting with suppliers for component parts. For example, an auto manufacturer must often contract with a tire supplier years before beginning the actual production of a new vehicle. Operations management can help OEMs determine how many suppliers to have of each part, where those suppliers should be globally, and how large the contract should be with each supplier.

Finally, when we think about the deployment of the COVID vaccine, we begin to understand how a failure to adequately embrace operations management can lead to sub-optimal results. As I write this, almost three in eight U.S. citizens have been fully vaccinated. But the underlying process could have been managed far better. One of our daughters lives in New York. On a recent visit to our home in Michigan, she contacted a friend who was able to secure a first dose of the vaccine for her at a pharmacy in Ohio. Weeks later she received her second vaccine in New York. How crazy is that? How many millions of person-hours were wasted as individuals searched far and wide for any clinic or pharmacy that could vaccinate them. Better use of operations management could have resulted in a far smoother rollout of the vaccines. Operations management could have helped determine how many vaccination sites to have in each city, where those sites should be, how many doses to allocate to each site, and how to prioritize the population of individuals seeking the vaccine. In fact, operations management has many applications in pandemic response planning [Akbun, Alumur, and Erenay, 2021, Dursunoğlu, Özdemir, and Dora, 2021, Ke and Zhao, 2021] and humanitarian logistics [Demir, Kara, and Sahinyazan, 2021, Rodríguez-Pereira et al., 2021, Sanci and Daskin, 2019, 2021].

In short, operations management deals with such topics as supply chain design, security screening, healthcare operations and medical decision-making, energy network design, telecommunications, inventory management, and a host of other areas. In fact, the application domains of operations management, sometimes referred to as operations research or management science, are limited only by our imagination.

1.2   WHAT DO THESE APPLICATIONS SHARE IN COMMON?

Given the broad range of applications of operations management, it is natural to ask what they all have in common. There are at least three common features of almost all applications of operations management:

1. a desire to improve, or in many cases to **optimize**, the performance of a system;

2. a recognition of the need to **tradeoff** competing demands placed on the system; and

3. an understanding that the conditions under which most systems operate are inherently **uncertain**.

To illustrate these facets of operations management, consider the decisions outlined in Table 1.1. These three features—**optimization, tradeoffs**, and **uncertainty**—will undergird the remainder of this book.

1.3   OPTIMIZATION, TRADEOFFS, AND UNCERTAINTY

We all want to do the best that we can. It is a natural human tendency or instinct. My parents instilled in me that very ethic: that I should always do my best. At the heart of operations management is the systematic study of ways in which processes can be improved, made better, or

Table 1.1: Sample operations management contexts

Context	Optimize	Tradeoff	Uncertainty
Distribution planning at Amazon	Number and location of fulfillment centers	Cost of having more fulfillment centers versus the speed with which packages can be delivered to customers	Future demand and shipping costs
TSA screening	Number of checkpoints in an airport and the configuration of each checkpoint	Cost of adding staff versus passenger delays	Daily travel demand and passenger arrival patterns
Cell phone tower deployment	Number and location of towers	Cost of towers versus customer service and regional coverage	Demand for call service as a function of the time-of-day
Contracting (snow removal or supplier)	Number of items to contract for (snow plowings or supplier capacity)	Cost of items contracted for versus lost sales or extra cost when demand exceeds contracted number	Demand for the service being provided (snowplowing) or the items being supplied (tires, for example)

optimized. The unofficial tagline of the Institute of Industrial and Systems Engineers is: "Engineers make things; industrial engineers make things better." This is not to say that an industrial engineer can design a better vaccine than a biomedical engineer, but that an industrial engineer trained in operations management can help design efficient and effective supply chains to deliver the vaccine from production plants to the arms of individuals needing immunization. Similarly, while industrial engineers never dream of replacing the deep medical knowledge of physicians caring for their patients, industrial engineers trained in operations management techniques can help design systems that facilitate equitable access to medical care by all members of the community. While deep knowledge of mechanical and electrical systems is needed to design wind turbines, operations management can help plan cost-effective maintenance schedules for these complex devices that enable wind farms to increase the time that they are operational [Byon, Ntaimo, and Ding, 2010].

Doing your best has limits, however. As a student, you can always devote another hour to studying for a calculus test, an exam in physics, or a quiz in probability. However, there are diminishing returns associated with each additional hour you spend studying any one subject as well as diminishing returns to studying overall. Other activities compete for the limited resource of your waking hours. In short, you must tradeoff an additional hour of studying physics against going for a walk with a friend or sharing a pizza with your roommate. Similarly, access to emergency medical services would surely be enhanced if each city in the United States doubled the number of ambulances it deployed. But that action comes at a cost. Perhaps a city would reduce the number of times it collected trash from neighborhoods. Would the overall health benefits of additional ambulances outweigh the health costs of less frequent garbage collection? Would more ambulances but fewer elementary school teachers be a good use of the public's money? There are clearly tradeoffs in just about every personal, corporate, or governmental decision.

Most decision-making is done in an environment of uncertainty about the future. One of my favorite sayings, attributed in various forms to Mark Twain, Niels Bohr, and Yogi Berra among others, is "It's tough to make predictions, especially about the future." When General Motors contracts with a supplier of seats for a new pickup truck, the demand for the truck two years from the contracting date is uncertain. Predictions at that stage can often be off by a factor of two or more. Similarly, exchange rates, which influence the ultimate cost of goods procured overseas, are highly variable and uncertain. Thus, the optimization of processes and the decisions about key tradeoffs must be made in the face of significant uncertainty about the future conditions under which the systems we are designing today will ultimately be operating [Sadghiani, 2018].

As a final note, in many contexts operations modeling is used to gain **insights** into the optimal operation of a system. The models themselves may not provide the exact solutions. For example, in work I did early in my career, I was part of a team [Eaton et al., 1985] that helped the city of Austin, TX establish bases for emergency medical vehicles. We ran tens of model runs varying key inputs. The final sites chosen by the city did not correspond to any single set of outputs from our models. Nevertheless, the modeling exercise was a success because each decision made by the city could be traced to some model run. For example, they may have not chosen to locate in one zone of the city recommended repeatedly by our model but instead located across the street in what was technically a different zone, but was an area in which the city already owned land. In this context, it is worth remembering the comment of George Box, one of the greatest statisticians of all time: "All models are wrong, but some are useful." Our goal is to develop useful models from which we can gain insights.

1.4   OUTLINE OF THE TEXT

Chapters 2–6 introduce and summarize probability and, to a much lesser degree, statistics. In Chapter 7, we discuss the fundamentals of optimization including inputs, decision variables, objectives, and constraints. Chapter 7 applies this framework to a classical problem in statistics, fitting a line to data. We apply optimization to a traditional problem in inventory theory illustrating a key tradeoff in Chapter 8. Chapter 9 merges optimization and uncertainty to examine a purchasing decision in the face of uncertainty.

Chapter 10 introduces one of the most common optimization modeling tools: linear programming. Chapter 11 applies linear programming to the problem of redistributing empty ride-sharing vehicles at the end of the day to positions at which they will be needed the next morning. Chapter 12 applies linear programming to the problem of assigning students to seminars to optimize average student satisfaction with their assigned seminars. In Chapter 13, we begin a discussion of location modeling. Here we will see the tradeoff between the cost of establishing a system and the level of service provided to customers of the system. Chapter 14 examines a location problem that could be used by a city to determine the locations of ambulance bases for its population. Chapter 15 examines the sort of problem that underlies some of Amazon's decision making about where to establish fulfillment centers or warehouses. Chapter 16 explores a key tradeoff between the objectives of Chapters 14 and 15.

Customers, or items to be processed, often wait for service. Unless you arrive at a particularly light travel time, you wait in two separate lines for TSA screening: to have your ticket and ID checked and to have your carryon luggage scanned. You wait to get an appointment with your doctor and you wait to see your doctor once you arrive at the clinic. Chapter 17 introduces queueing theory, the mathematical theory of waiting lines. In Chapters 18–20, we explore progressively more complicated queues and derive principles about how the waiting time depends on the (1) utilization of the system, (2) the variability of the service times, and (3) the size of the system as measured by the number of service stations. Chapter 21 moves from analytic queueing theory to more complex models that require computer simulation and numerical methods. Chapter 22 introduces decision theory, which is another approach to making decisions in the face of uncertainty. In its simplest form, decision theory could be used to help you decide if you should take an umbrella to work tomorrow morning and in more complex situations, it can be used by government agencies to determine whether or not to lock down a city or state during a pandemic. We outline other areas of operations management in Chapter 23.

1.5   REFERENCES

E. Akbun, S. A. Alumur, F. S. Erenay, Determining the optimal COVID-19 testing centre locations and capacities considering the disease dynamics and target populations, *International Symposium on Locational Decisions, ISOLDE*, University of Wuppertal, Germany, July 5–9, 2021. 2

E. Byon, L. Ntaimo, and Y. Ding, Optimal maintenance strategies for wind turbine systems under stochastic weather conditions, *IEEE Transactions on Reliability*, 59(2):393–404, 2010. DOI: 10.1109/tr.2010.2046804. 3

Ş. M. Demir, B. Y. Kara, and F. Sahinyazan, Districting for integration of Syrian refugees into the Turkish education system, *International Symposium on Locational Decisions, ISOLDE*, University of Wuppertal, Germany, July 5–9, 2021. 2

Ç. Dursunoğlu, I. Özdemir, B. Y. Kara, and M. Dora, Logistics of half-mobile testing booths for COVID-19, *International Symposium on Locational Decisions, ISOLDE*, University of Wuppertal, Germany, July 5–9, 2021. 2

D. Eaton, M. S. Daskin, D. Simmons, B. Bulloch, and G. Jansma, Determining emergency medical service vehicle deployment in Austin, Texas, *Interfaces*, 15(1):96–108, 1985. DOI: 10.1287/inte.15.1.96. 4

G. Y. Ke and J. Zhao, Infectious waste management with demand uncertainty during a pandemic, *International Symposium on Locational Decisions, ISOLDE*, University of Wuppertal, Germany, July 5–9, 2021. 2

J. Rodríguez-Pereira, G. Laporte, M.-È. Rancourt, and S. Silvestri, Location of water tabs and design of a water distribution network in Nepal, *International Symposium on Locational Decisions, ISOLDE*, University of Wuppertal, Germany, July 5–9, 2021. 2

E. Sanci and M. S. Daskin, An integer L-shaped algorithm for the integrated location and network restoration problem in disaster relief, *Transportation Research Part B: Methodological*, 145:152–184, 2019. DOI: 10.1016/j.trb.2021.01.005. 2

E. Sanci and M. S. Daskin, Integrated location and network restoration decisions in relief networks under uncertainty, *European Journal of Operational Research*, 279(2):335–350, 2021. DOI: 10.1016/j.ejor.2019.06.012. 2

N. Salehi Sadghiani, Models for flexible supply chains, Ph.D. Dissertation, Industrial and Operations Engineering Department, University of Michigan, Ann Arbor, MI, 2018. 4

C H A P T E R   2

Fundamentals of Uncertainty Analysis

2.1 PROBABILITY AND STATISTICS

Chapter 1 indicated that one of the pillars of operations management is understanding, managing, and mitigating the effects of uncertainty or randomness. In general, though not always, randomness is the enemy of operations management. In this chapter we discuss and contrast two related disciplines: probability and statistics.

In probability and statistics, we think about *experiments*. An experiment is not just what happens in a laboratory; rather, we think about an experiment as any action or process whose outcome is uncertain. Thus, the number of patients arriving at a clinic each hour is an experiment. The time you spend waiting for a bus is also an experiment.

We distinguish between a *population* (or the *sample space*) and a *sample* from the population. The population is the set of all possible outcomes of an experiment. Thus, if we are modeling the number of people arriving at a clinic each hour, the population is the set of all non-negative integers. The likelihood of getting a very large number of patients (e.g., more than 20) is generally very small, but conceivably you could get a large number. If we are modeling the time you wait for a bus, the population is the set of all non-negative real numbers, since you could wait a fraction of an hour.

A sample is a realization or set of realizations from the population. For example, if the population represents the number of patient arrivals each hour and the true average arrival rate is three per hour, we might observe values as shown in Table 2.1. In some hours, we get no arrivals, and in some cases as many as seven patients arrive. Also, note that the daily averages are not all equal to three even though the population has an average of three.

Figure 2.1 illustrates what probability and statistics do with respect to the population and a sample. In particular, if we know everything there is to know about the population, probability lets us say something about what samples from the population look like. For example, if we knew that 30% of the students at the University of Michigan were 19 years old or younger, probability would tell us the likelihood of finding 32 or more such students in a sample of 100 students. Conversely, if we have a sample, statistics lets us make statements about what the unknown population is likely to be. Thus, if we sampled 100 students and found 33 students were 19 or younger, statistics would allow us to say whether or not the interval below 30% is likely to include the true fraction. (That is worded awkwardly for technical reasons beyond our scope.)

Table 2.1: Sample of the number of patients arriving each hour

Hour	Monday	Tuesday	Wednesday	Thursday	Friday
1	1	5	3	2	7
2	0	4	4	3	5
3	3	3	6	4	3
4	1	2	4	1	4
5	5	0	1	4	4
6	6	2	4	3	3
7	3	3	3	2	5
8	2	5	2	1	3
9	3	1	3	5	3
10	3	4	4	1	0
11	2	0	2	1	3
12	3	7	4	3	3
Daily average	2.67	3.00	3.33	2.50	3.58

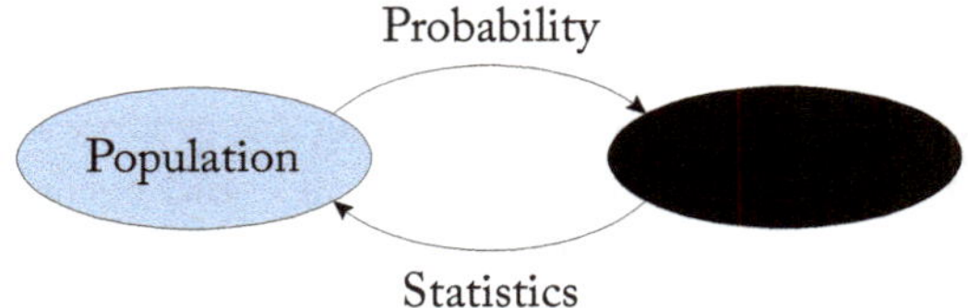

Figure 2.1: Population, sample, probability, and statistics.

2.2   EVENTS AND AXIOMS OF PROBABILITY

An *event* is a collection of outcomes of an experiment. The *union* of two events A and B, denoted $A \cup B$, is the set of all outcomes in either of the two events. The *intersection* of two events A and B, denoted $A \cap B$, is the set of all outcomes in both events. Two events are *mutually exclusive* if the intersection of the two events is the empty set. Similarly, a set of events is *collectively exhaustive* if the union of the events in the set is the same as the sample space.

Probabilities measure the likelihood of an event happening. A probability of 0 means the event will never happen and a probability of 1 means the event will always happen. In particular, we have the following three key axioms of probability.

Table 2.2: Joint and marginal probabilities in healthcare testing

<table>
<tr><td rowspan="2"></td><td rowspan="2"></td><td colspan="2" align="center">Health Condition</td><td rowspan="2"></td></tr>
<tr><td align="center">Sick</td><td align="center">Not Sick</td></tr>
<tr><td rowspan="2" align="center">Test
Condition</td><td align="center">Positive</td><td align="center">True Pos</td><td align="center">False Pos</td><td align="center">Marginal Test
Positive</td></tr>
<tr><td align="center">Negative</td><td align="center">True Neg</td><td align="center">False Neg</td><td align="center">Marginal Test
Negative</td></tr>
<tr><td></td><td></td><td align="center">Marginal
Positive</td><td align="center">Marginal
Negative</td><td></td></tr>
</table>

1. For any event A, $p(A) \geq 0$, where $p(A)$ denotes the probability that event A occurs.

2. $p(S) = 1$ where S represents the sample space or population.

3. If $A_1, A_2, \ldots, A_k$ are mutually exclusive events, then $p(A_1 \cup A_2 \cup \ldots \cup A_k) = \sum_{n=1}^{k} p(A_n)$. If $A_1, A_2, \ldots$ is an infinite collection of mutually exclusive events, then $p(A_1 \cup A_2 \cup \ldots) = \sum_{n=1}^{\infty} p(A_n)$.

The axioms have a number of key implications. First, they imply that $p(A) = 1 - p(A^c)$ where A^c is the *compliment* of event A, or the event that A does not happen. They imply that if A and B are mutually exclusive events, then $p(A \cap B) = 0$. They also imply that $p(A \cup B) = p(A) + p(B) - p(A \cap B)$.

2.3   JOINT, MARGINAL, AND CONDITIONAL PROBABILITIES AND BAYES' THEOREM

Finally, we turn our attention to *joint*, *marginal*, and *conditional* probabilities. These are perhaps best illustrated by medical testing. Consider testing for a particular disease like COVID-19. Some fraction of the population actually has the disease and the rest of the population does not have the disease. Among those who are tested for the disease, some tests will come back positive and some will come back negative. Table 2.2 illustrates all of the possibilities. At the top of the table, in red, are the true health conditions: either a patient is sick or the patient is not sick. On the left-hand side of the table, in blue, are the test conditions: either a test comes back positive or it comes back negative. This means that there are four possible outcomes as shown with the pink shading: (1) a patient is sick and the test comes back positive, which we call a true positive; (2) a patient is not sick, but the test comes back positive, which is a false positive; (3) a patient is sick, but the test comes back negative, which is a false negative; or (4) the patient is not sick and the test comes back negative, which is a true negative.

Table 2.3: Sample joint and marginal probabilities in healthcare testing

		Health Condition		
		Sick	Not Sick	
Test	Test Pos	0.245	0.06	**0.305**
Condition	Test Neg	0.005	0.69	**0.695**
		0.25	**0.75**	

When we fill in the table with numbers, the values shown with the pink shading are referred to as *joint* probabilities since they refer to two events happening simultaneously, e.g., the event that a patient is truly sick and the test comes back positive as shown in the top left-hand side of the pink shaded region. The sums of the row or column values are the *marginal* probabilities and they indicate the fraction of the population that is truly sick (the marginal positive cell shown with blue shading) or the probability that a test comes back positive (the marginal test positive cell shown with green shading).

It is worth noting that the sum of all of the pink shaded numbers must equal 1. Similarly, the sum of each of the two sets of marginal probabilities (shown in green or blue shading) must be 1 as well.

Table 2.3 shows sample values for a disease with a prevalence of 25%, a test *sensitivity* of 98%, and a test *specificity* of 92%. What do all these terms mean? The *prevalence* is the fraction of the population that truly has the disease. This is simply the marginal probability of someone being sick. The test *sensitivity* is the ratio of the true positive cases divided by the fraction of the population that is positive. In this case that is 0.245/0.25 or 0.98. The test *specificity* is the ratio of the true negative cases divided by the fraction of the population that is not sick, or 0.69/0.75 or 0.92.

Bayes Theorem tells us that

$$p(A \text{ occurs given } B \text{ has happened}) = p(A|B) = \frac{A \cap B}{p(B)}.$$

We can use Bayes Theorem to compute the probability that you are sick *given* that the test came back positive. We simply let A be the event that you are sick and B be the event that the test came back positive. Thus, we would have:

$$p(\text{Sick}|\text{Test Positive}) = \frac{p(\text{Sick} \cap \text{Test Positive})}{p(\text{Test Positive})} = \frac{0.245}{0.305} = 0.80.$$

Therefore, while a positive test is a cause for concern in this case, it is not the same as a definitive diagnosis. Most doctors would tell you that you should have other, perhaps more invasive, tests to confirm the diagnosis.

2.4 INDEPENDENT EVENTS

Finally, it is important to define what we mean by *independent* events. Two events, A and B, are independent if knowledge that one event occurred does not impact the probability that the other occurred. More rigorously, events A and B are independent if $p(A|B) = p(A)$ and $p(B|A) = p(B)$. Clearly, two mutually exclusive events, cannot be independent since knowing that event A occurred, for example, will tell you that event B could not have occurred and vice versa. Finally, we note that if events A and B are independent, then $p(A \cap B) = p(A)p(B)$. In other words, if A and B are independent, then the probability of both events occurring is the product of the probabilities that each event occurs.

C H A P T E R   3

Intuition About Probability

3.1   INTRODUCTORY COMMENTS

Probability is not a simple concept. For many of us, our exposure to probability is limited to card games and other games of chance like Monopoly or Backgammon. Since very few of us will become professional card players or Monopoly experts, we will generally avoid these examples in this text. Instead, we will focus on other areas of life in which uncertainty plays a big role.

In this chapter, we will introduce two problems that will reinforce how counter-intuitive some issues are in probability. The first is the so-called *birthday problem* and the second a problem related to forming a triangle from three pieces of wood.

3.2   THE BIRTHDAY PROBLEM

You are having a number of friends over for a party. How likely is it that at least two of you were born on the same day of the year? (Throughout this problem, we will ignore February 29, since less than one in every 1,450 people was born on that date which occurs about once every four years.) How does this probability depend on how big the party is? How many people do you need to have at the party so that there is at least a 50:50 chance of having two people with the same birthday?

STOP READING FOR A MINUTE AND USE YOUR INTUITION TO ANSWER THE LAST QUESTION ABOUT HOW MANY PEOPLE ARE NEEDED AT THE PARTY FOR THERE TO BE A 50:50 CHANCE OF TWO OR MORE PEOPLE SHARING A BIRTHDAY.

Many people will say that you need 366 people in the room. That is the correct answer if we want to be absolutely sure that at least two people share the same birthday. Others might say at least 100 people. The correct answer is 23. Once you have 23 people at the party, the probability that everyone has a different birthday is less than 0.5.

To see how to analyze this problem, we will compute the probability that all N people at a party have different birthdays. Let us start with one person at the party. Clearly, at that point, you cannot have two people with the same birthday. If you have two people at the party, the probability that the second person has a birthday that is different from that of the first person is 364/365. If we add a third person, that person can have one of the remaining 363 birthdays in the year if all three are to have different birthdays. Thus, the probability that all three have different

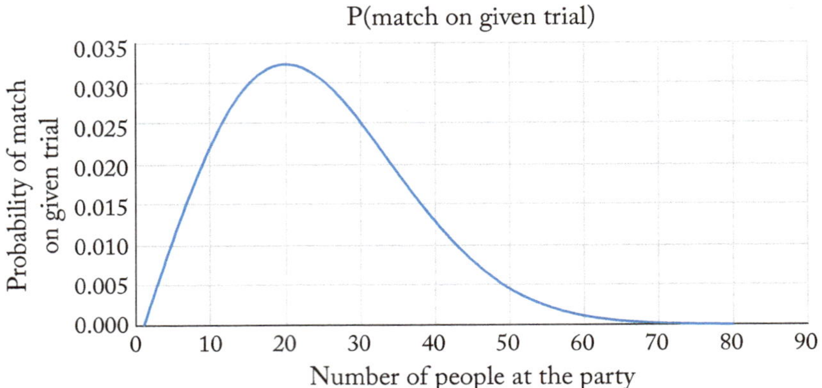

Figure 3.1: Probability first match occurs with given number of people at the party.

birthdays is $1 \cdot \frac{364}{365} \cdot \frac{363}{365} \approx 0.986$. If we have 22 people in the room the probability that all have different birthdays is $1 \cdot \frac{364}{365} \cdot \frac{363}{365} \ldots \frac{344}{365} \approx 0.524$. With 23 people in the room, the probability that all 23 will have different birthdays is $1 \cdot \frac{364}{365} \cdot \frac{363}{365} \ldots \frac{343}{365} \approx 0.493$.

Now let us consider an experiment in which we keep adding people to the party until we have a birthday match. We then record the number of people at the party when we obtained the match. We repeat this experiment many times. It turns out that the average number of people in these parties is 24.6. This highlights that the *average* of a distribution (24.6 in this case) is not the same as the *median* or value of the random variable, such that half or more of the probability is to the left of the median and half or more is to the right of the median (23 in this case).

Figure 3.1 plots the probability that the first match occurs when we add the given person to the party. Thus, the *mode* of the distribution, the random number with the highest probability, is 20. When we add the 20th person to the party, the probability that there will be a match at that point exactly is (only) about 0.032. Figure 3.2 shows the cumulative distribution of the probability distribution. This shows that the median of the distribution is when we have 23 people at the party.

3.3   THE YARDSTICK PROBLEM

Now we will turn to a different problem. Let's take a yardstick that is 36 inches long. Pick two positions at random along the yardstick. Cut the yardstick at those points. This will result in three pieces of wood. Yes, I did that in class at Northwestern University on a number of occasions. What is the probability that you can make a triangle out of the three pieces?

STOP READING FOR A MINUTE AND USE YOUR INTUITION TO ANSWER THIS QUESTION.

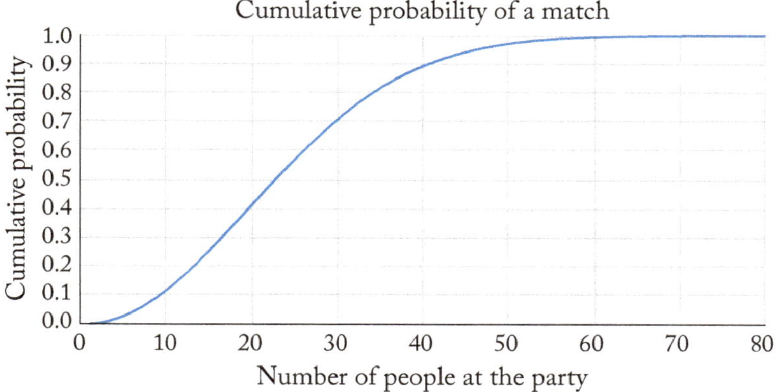

Figure 3.2: Probability of a match with given number of people or fewer.

Many people will say that the probability is 1. After all, there are three pieces of wood! But let's think about this a bit. Suppose the first cut is at 1 inch and the second is at 34.5 inches. This will leave us with three pieces of wood that are 1 inch, 33.5 inches, and 1.5 inches long. Since the sum of the two shorter sides (2.5 inches) is less than the length of the longer side, we cannot make a triangle out of these three pieces.

Figure 3.3 shows the joint sample space for this problem. There are two dimensions to the graph corresponding to where each of the two cuts appears. This is the *joint sample space* for this problem. Most of the graph is red, indicating that these are regions in which you cannot make a triangle. Since all points in the sample space are equally likely, the probability you can make a triangle when you cut the yardstick in this way is the area of the two green triangles divided by the area of the entire samples space, or only 0.25.

These two examples should demonstrate that, for most of us, our intuition about probability is faulty at best. The next few chapters outline some standard distributions that are used in much of our probabilistic work in operations management.

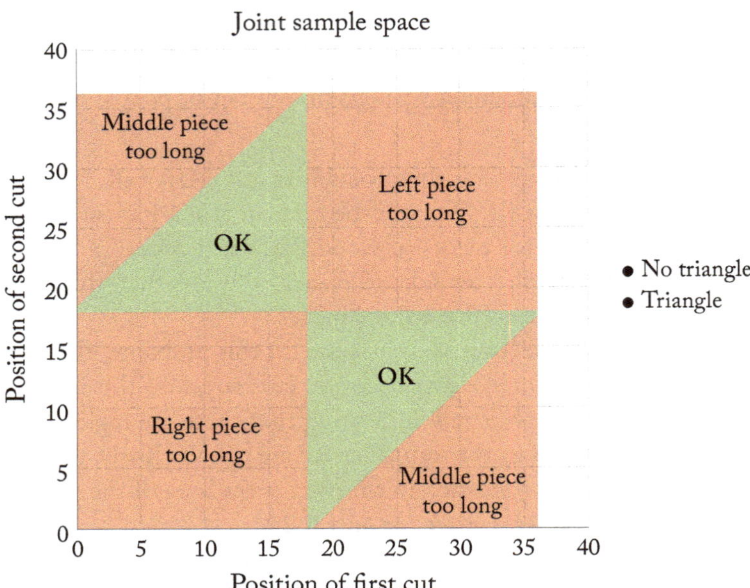

Figure 3.3: Joint sample space for the yardstick problem.

CHAPTER 4

Discrete Probability Distributions

4.1 DISCRETE AND CONTINUOUS RANDOM VARIABLES

In this and the next chapter, we introduce some often-used probability distributions. Random variables generally come in one of two flavors: discrete or continuous, though as we will see in our look at queueing theory some random variables can be both discrete and continuous. By discrete, we mean that the random variable can only take on specified or "discrete" values, though the number of such values may be infinite. One way to think about discrete random variables is that they can be represented by integer values. Continuous random variables, on the other hand, can take on an infinite set of values and can be thought of as being represented by real numbers. In this chapter we will examine discrete random variables. Chapter 5 will introduce continuous random variables.

4.2 SOME INTRODUCTORY NOTATION

We will begin with some simple notation. We let X denote a generic random variable. We use the notation $p(X = x)$ or more simply $p(x)$ to denote the probability that the random variable X takes on a specific value x. For example, X might represent the number of patients to arrive at a walk-in clinic between noon and 1:00 pm on a particular Saturday. This is clearly a random variable, since we do not know *a priori* how many patients will arrive. We might have $p(X = 0) = 0.05$ or more simply $p(0) = 0.05$, meaning that there is a 5% chance that there will be no patients arriving at the clinic during this period of time. Similarly, we might have $p(X = 1) = 0.15$ or $p(1) = 0.15$. Clearly, we require $\sum_{x=0}^{\infty} p(X = x) = 1$ since (a) there cannot be a negative number of arrivals and (b) the sum of all possible numbers of arrivals must equal 1. This is equivalent to axiom 2 of Chapter 2 which stated, $p(S) = 1$, where S is the sample space.

We are often interested not only in the probability that a particular event happens (e.g., that there is exactly one arrival during a given hour), but we are also interested in the probability that there are x or fewer events. We will use the notation $P(X \leq x)$ or simply $P(x)$ to denote this probability. Note that we are using an upper case P to denote this probability. This is called a *cumulative probability*. Using the values above, we would have $P(X \leq 0) = P(0) = 0.05$ and $P(X \leq 1) = P(1) = 0.2$. Also, for any random variable, $P(X \leq \infty) = P(\infty) = 1$.

4.3   BERNOULLI TRIALS

We will begin our discussion of discrete probability distributions by introducing *Bernoulli trials*. A Bernoulli trial is an experiment that can have two outcomes. Examples include: (1) true/false; (2) won/lost; (3) has an earned undergraduate degree/does not have such a degree; (4) has tested positive for COVID /has not tested positive for COVID; and (5) was convicted of a DUI offense/was not convicted of a DUI offense. The last three examples reflect possible outcomes of sampling individuals in a population.

We will denote one of these outcomes—true, won, has an earned undergraduate degree, has tested positive for COVID, and was convicted of a DUI offense—as being a *success* and the other outcome, in each case, as being a *failure*. The designation of one as a success and one as a failure is purely arbitrary. We let the probability of a success be denoted by q in each case.

4.4   THE BINOMIAL DISTRIBUTION

We can then ask, if we sample N Bernoulli trials independently (meaning that each trial does not depend in any way on any previous or successive trial), what is the distribution of the number of successes out of the N trials. It turns out that this is given by the *Binomial* distribution with the following form:

$$P(X = n) = \binom{N}{n} q^n (1 - q)^{N-n} \quad n = 0, 1, \ldots, N.$$

Again, X is the random variable and n is a particular value of the random variable. To have n successes means that we would have $N - n$ failures. The probability of n successes is q^n and the probability of $N - n$ failures $(1 - q)^{N-n}$. Finally, there are

$$\binom{N}{n} = \frac{N!}{n!(N - n)!}$$

possible ways of getting n successes and $N - n$ failures. (Note that $n! = n(n-1)\ldots 1$ and we define $0! = 1$.) For example, we could have n successes followed by the $N - n$ failures; we could have $N - n$ failures followed by n successes; or we could have a (potentially) large number of other ways of getting n successes and $N - n$ failures out of N trials.

For example, suppose your favorite team—undoubtedly the University of Michigan Wolverines—plays 12 games in a season and the probability of their winning any game is 0.6. Furthermore, assume that the probability of winning is the same each week and that the probability of winning in one week does not depend on how well they did in the previous week's

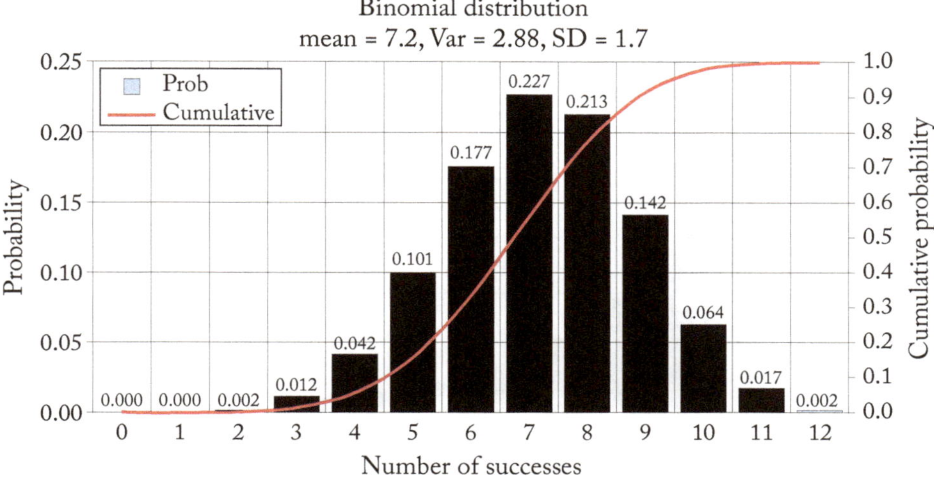

Figure 4.1: Sample binomial distribution.

game. Under these conditions, Figure 4.1 plots the probability mass function of the number of wins during the season in blue. The red line shows the cumulative distribution. While this is plotted as a smooth line, it would be better to think of it as a step function. Thus, the cumulative probability of 6 or fewer wins is 0.335, meaning that in about 1/3 of many such seasons, the team would amass 6 or fewer wins out of 12 games.

In Chapter 6, we will return to the concepts of the mean, variance and standard deviation shown in the title of the figure. Suffice it to say that the mean is simply the average number of games we would expect the team to win. It is important at this point, however, to stress that while the distribution is defined only for integer-valued outcomes $(0, 1, \ldots, 12$ wins), the average can be a fraction. One way to get a feel for how this can happen is to think about the average fertility rate of women by country. The World Factbook lists the fertility rate, or the average number of births per woman, by country.[1] It ranges from a high of 6.91 to a low of 1.07. The United States stands at 1.84. The population-weighted global average is 2.40. Clearly, any single woman will have an integer number of children. However, when we take an average over all women in a country, it should be easy to see that we may get fractional values. Similarly, the mean above represents the theoretical average number of wins that such a team would accumulate over each of many identical seasons.

---

[1] https://www.cia.gov/the-world-factbook/field/total-fertility-rate/country-comparison

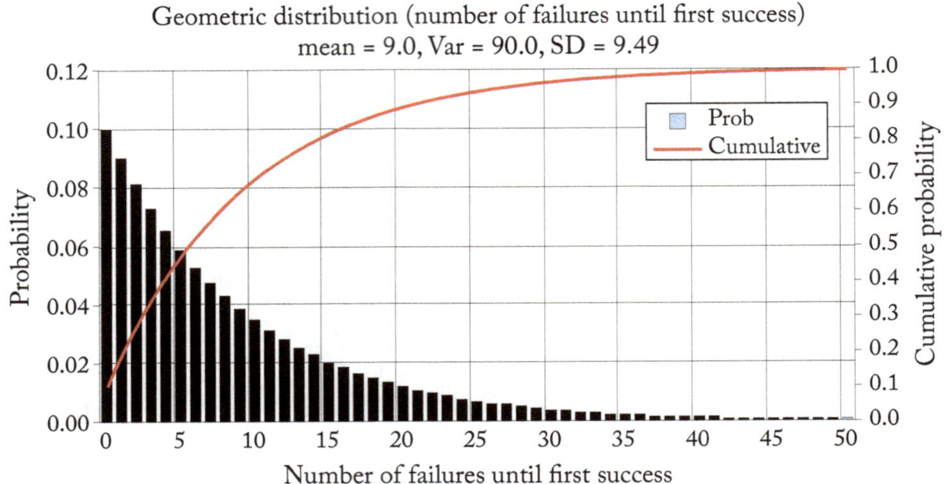

Figure 4.2: Sample geometric distribution.

4.5   THE GEOMETRIC DISTRIBUTION

In addition to asking about the number of wins a team will have in a season, we can ask what is the distribution of the number of games a team will lose before their next win. It could be as little as 0, and this happens with probability q if q is the probability of winning any game. If they lose the next game, which happens with probability $(1 - q)$ and win the next game, then they will lose one game before their first win. This happens with probability $(1 - q)q$. Similarly, the probability that the team will lose $n - 1$ games before their first win is $(1 - q)^{n-1}q$. In other words, if Y is a random variable indicating the number of games the team will lose before the first win, we have

$$p(Y = n) = (1 - q)^n q \quad n = 0, 1, \ldots .$$

Observe first that the number of possible losses for this distribution extends to infinity. Second, each term is equal to the preceding term multiplied by $(1 - q)$. Thus, each term is smaller than the preceding term by a multiplicative factor of $(1 - q)$. The distribution of the number of failures until the first success is called a *geometric* distribution.

Suppose, for example, that 10% of a population tests positive for a particular disease. Figure 4.2 plots the distribution of the number of people who would test negative before we get the first person who tests positive. With probability 0.1, the first person tested will be positive and so there will be 0 failures or 0 people testing negative before getting the first positive test. With probability 0.09, the first person will be negative and the second person will be positive. Again, the red line shows the cumulative distribution and, again, it might be better to think of this as a step function. On average, we will test nine people before getting the first positive case on the 10th trial.

The geometric distribution—like its continuous counterpart, the exponential distribution, covered in Chapter 5—has the *memoryless* property. Our intuition suggests that if, on average it takes nine failures before the first success and we have already tested five people all of whom are negative or failures, then we should only have to test four more on average before getting the first success on the 5th trial after the first five negative tests. Unfortunately, this intuition is **wrong**. To see this analytically, let us compute

$$p \, (n \text{ more failures before the first success GIVEN } m \text{ failures so far}) =$$

$$\frac{p(n + m \text{ failures before the first success})}{p(m \text{ failures so far})} =$$

$$\frac{(1 - q)^{m+n}q}{(1 - q)^m} = (1 - q)^n q.$$

But this is nothing more than the original geometric distribution. In other words, the system does not remember what happened before. The expected number of additional people we would have to test before we find the first positive person is still nine, even though we have already seen five successive people who tested negative. Another way to think about this is to think about tossing a coin. On average, we would expect to get heads 50% of the time. This means that on average we would get one tails before the first heads. Of course, we could get 0 tails before getting heads on the first coin toss. Now, suppose we have had tails three times in a row (which happens with probability 0.125). Our poor coin cannot remember that it had three tails in a row and so it does not know that we are "due" to get heads. In fact, in this context, being "due" for heads has no meaning.

4.6   THE POISSON DISTRIBUTION

Finally, we can ask what happens to the Binomial distribution if we keep Np constant and equal to λ, but let N get very large (or approach infinity). This distribution is the *Poisson* distribution and is given by:

$$p(X = n) = \frac{\lambda^n e^{-\lambda}}{n!} \quad n = 0, 1, \dots.$$

Again, the range of this distribution extends to infinity. Figure 4.3 plots a sample Poisson distribution with a $Np = \lambda = 7.2$. Figure 4.4 compares this Poisson distribution with the Binomial distribution with $N = 20$ on the left and $N = 100$ on the right. On the left, the agreement between the Poisson distribution in red and the Binomial distribution in green is poor. On the right, the agreement between the Poisson distribution (red) and the Binomial distribution (blue) is quite good, confirming that as we hold Np constant and increase N, the Binomial distribution approaches the Poisson distribution with $Np = \lambda$.

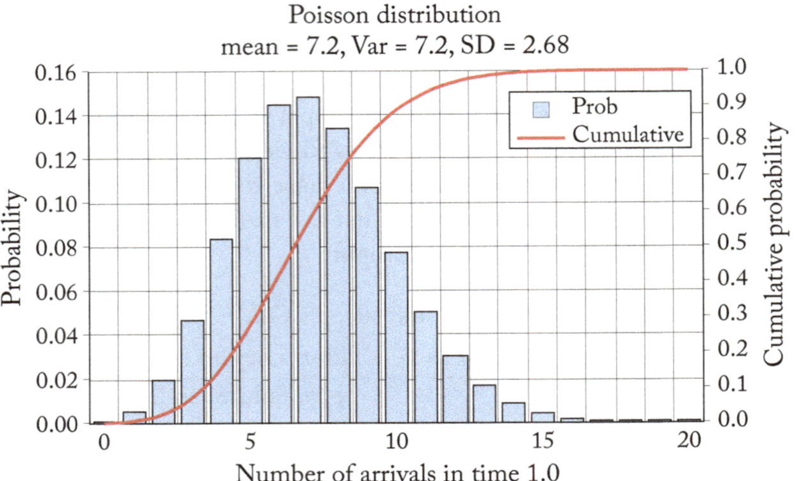

Figure 4.3: Sample Poisson distribution.

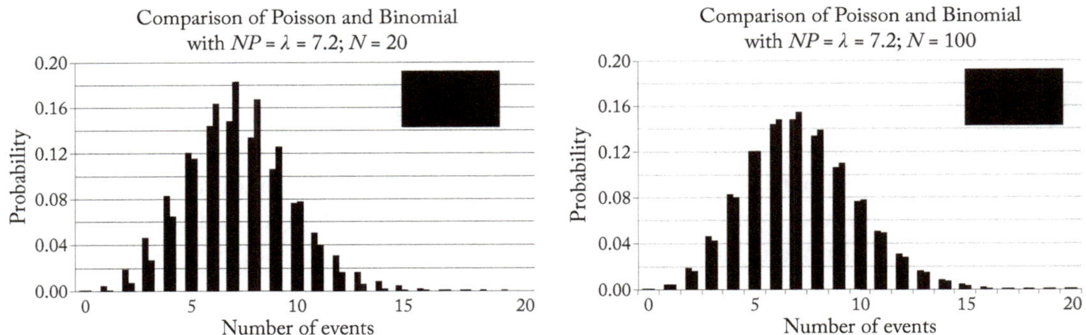

Figure 4.4: Comparison of Poisson and binomial distributions with $\lambda = 7.2$, $N = 20$, and $N = 100$.

The Poisson distribution is often used to model the arrival of customers at a facility. In this case, we can think of λ as the arrival *rate* of customers per unit time. During an interval of duration τ, the expected number of arrivals will be $\lambda\tau$ and the Poisson distribution will be given by

$$p[X(\tau) = n] = \frac{(\lambda\tau)^n e^{-\lambda\tau}}{n!} \qquad n = 0, 1, \ldots,$$

where $X(\tau)$ is the random variable denoting the number of arrivals during a period of duration τ.

C H A P T E R   5

Continuous Probability Distributions

5.1   BASIC CONCEPTS AND NOTATION

In the preceding chapter, we indicated that there are two broad classes of random variables: discrete random variables and continuous random variables. (Sorry, the opposite of a discrete random variable is not an indiscrete random variable.) Chapter 4 focused on discrete random variables; here we focus on continuous random variables. Recall that a continuous random variable is one that can take on any real value between some lower bound (potentially negative infinity) and some upper bound (again, potentially positive infinity).

For continuous random variables, instead of a *probability mass* function, we have a *probability density* function. For example, suppose T is a random variable denoting the time between customer arrivals at a coffee shop. $f_T(t)$ would give the density of the probability function at t. It is important to realize, that this is *not* the probability that the time is equal to t. In fact, it does not make sense to talk about the probability that the time is exactly equal to t or any other value such as a. To see this, note that we would have to evaluate $\int_a^a f_T(t)$ to find the probability that the time was *exactly* equal to a, but this integral is 0. We can think of the density function as giving the thickness of peanut butter on a slice of bread. While it makes sense to talk about how thick the peanut butter is at any point, it does not make sense to talk about how much peanut butter is at a particular (infinitesimally small) point, since that is 0, no matter how thick the peanut butter is at that point.

However, it does make sense to talk about the probability that a random variable takes on a value less than or equal to some value a. For example, we can ask what is the probability that the time between customer arrivals is less than a. This is given by $F_T(a) = \int_{-\infty}^a f_T(x)dx = \int_0^a f_T(x)dx$ (since time must be a non-negative random variable). This is called the *cumulative distribution* function. Loosely speaking, this is akin to asking how much peanut butter is to the left of some line on the piece of bread.

As before, we are using f to denote the density function and F to denote the cumulative distribution function.

5.2   UNIFORM RANDOM VARIABLES

We will begin with the simplest of continuous random variables, the *uniform* random variable. A *standard uniform* random variable is a real number selected at random between 0 and 1. The EXCEL function RAND() returns such a number each time it is invoked. Many calculators (or calculator apps) also have such a function.

The distribution of a *standard uniform random variable* is given by:

$$f_X(x) = \begin{cases} 1 & 0 \le x \le 1 \\ 0 & \text{elsewhere.} \end{cases}$$

The cumulative distribution function for a standard uniform random variable is given by:

$$F_X(x) = \begin{cases} 0 & x < 0 \\ x & 0 \le x \le 1 \\ 1 & x > 1. \end{cases}$$

We can also have a uniform random variable that lies between a lower limit of A and an upper limit of B. In this case, the density function is given by:

$$f_X(x) = \begin{cases} \dfrac{1}{B-A} & A \le x \le B \\ 0 & \text{elsewhere.} \end{cases}$$

In this case, the cumulative distribution is given by

$$F_X(x) = \begin{cases} 0 & x < A \\ \dfrac{x-A}{B-A} & A \le x \le B \\ 1 & x > B. \end{cases}$$

Note that if $B - A < 1$, the density function is greater than 1. This is fine, since the density function will still integrate to 1 when integrated over the entire range of the random variable. This just means that the peanut butter is very thick, but over a very small area.

Figure 5.1 plots a uniform random variable with $A = 5$ and $B = 30$. The density function is in blue and the cumulative distribution is in red. The mean or average is equal to 17.5 and is shown as a light green dashed line.

5.3   THE EXPONENTIAL DISTRIBUTION

If the number of arrivals during a particular time period follows a Poisson distribution with rate λ, then the distribution of time between arrivals follows an exponential distribution given by:

$$f_T(t) = \lambda e^{-\lambda t} \quad t \ge 0.$$

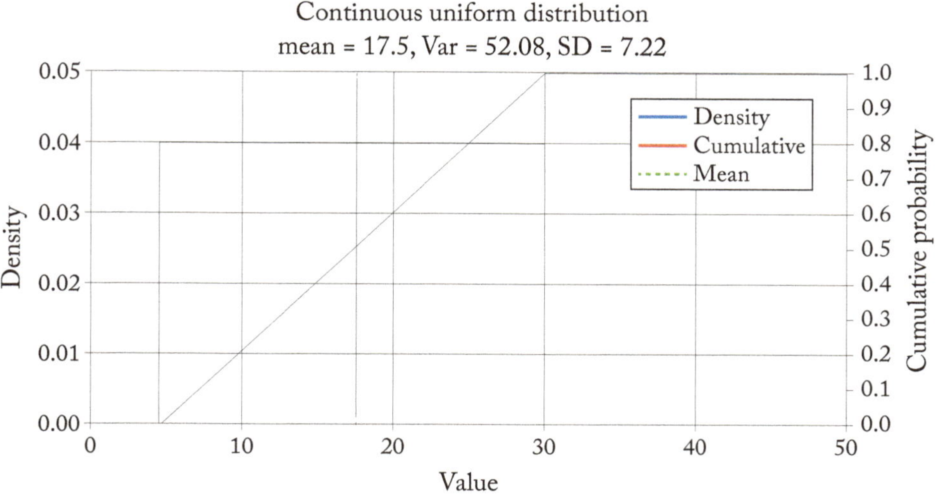

Figure 5.1: Uniform density function with $A = 5$ and $B = 30$.

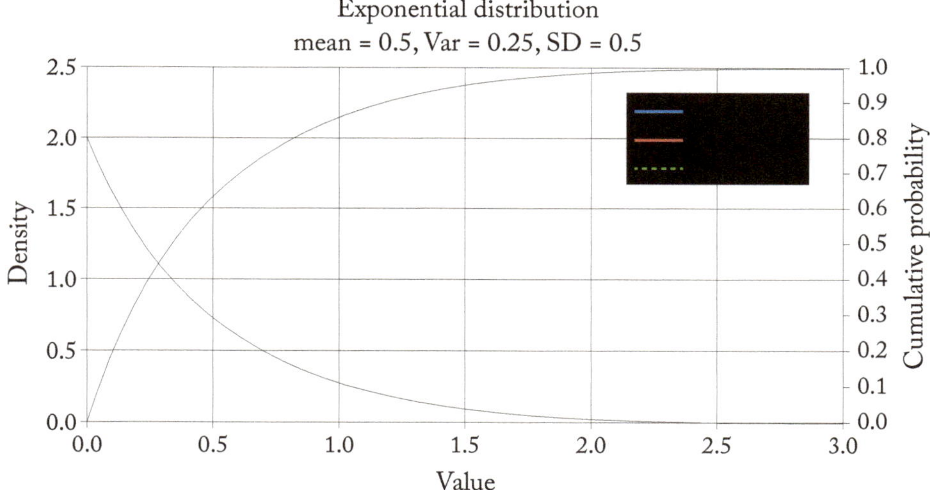

Figure 5.2: Exponential distribution with $\lambda = 2$.

The cumulative exponential distribution is given by

$$F_T(t) = 1 - e^{-\lambda t} \quad t \geq 0.$$

Figure 5.2 plots the exponential density function (blue) and the cumulative exponential distribution (red) when $\lambda = 2$. Note that the density function exceeds 1.0 for small values of time (in fact for all times $t \le \frac{-\ln(0.5)}{2} \approx 0.347$, the density function exceeds 1. However, the density function clearly integrates to 1.0.

Like its discrete counterpart, the geometric distribution, the exponential distribution is memoryless. What this means is that if T is a random variable that is exponentially distributed, then:

$$P(T \ge \gamma + \tau | T \ge \gamma) = \frac{P(T \ge \gamma + \tau \text{ AND } T \ge \gamma)}{P(T \ge \gamma)} = \frac{e^{-\lambda}(\gamma + \tau)}{e^{-\lambda \gamma}} = e^{-\lambda \tau}.$$

But this is nothing more than the probability that an exponential random variable with parameter λ is greater than τ. What this means is, for example, if the time between bus arrivals at a stop is exponentially distributed, then knowledge that it has been at least 15 minutes since the last bus arrival and that the average time between bus arrivals is 20 minutes, simply means that the distribution of the *additional* time until a bus arrives is still exponentially distributed with a mean of 20 minutes. In other words, there is no reason to ask someone at the stop how long they have been waiting, unless, of course, you just want to strike up a conversation with that person. When we get to queueing in Chapters 17–21, this will prove to be a very useful property.

5.4   THE ERLANG-k DISTRIBUTION

Another useful distribution is the *Erlang-k* distribution. This is the distribution of the *sum* of k independent, identically distributed, exponential random variables, each with parameter λk. If S is a random variable with an Erlang-k distribution, then the density function of S is given by:

$$f_S(s) = \frac{\lambda k (\lambda k s)^{k-1} e^{-\lambda k s}}{(k-1)!} \quad s \ge 0.$$

The cumulative Erlang-k distribution is given by:

$$F_S(s) = 1 - \sum_{n=0}^{k-1} \frac{(\lambda k s)^n e^{-\lambda k s}}{n!} \quad s \ge 0.$$

Note that the term in the summation above, is nothing more than a Poisson probability with parameter $\lambda k s$. Thus, there is an intimate link between the discrete Poisson distribution, the exponential distribution and the Erlang-k distribution. Also notice that for $k = 1$, the Erlang-k distribution is simply the exponential distribution as expected.

Figure 5.3 plots the Erlang-k distribution for five different values of the parameter k. All five distributions have a mean of 0.5. As k increases, the distribution collapses around the mean value. Figure 5.4 plots the corresponding cumulative distributions. As k increases, the cumulative distribution begins to look like an S-shaped curve. For very large values of k, the cumulative distribution will look like a step function, jumping from 0 to 1 very quickly.

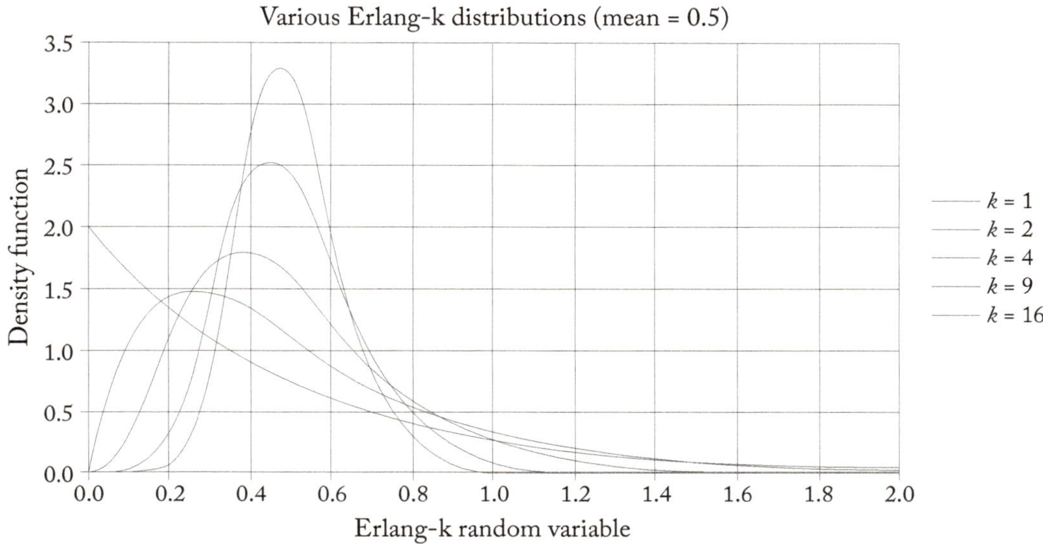

Figure 5.3: Various Erlang-*k* density functions.

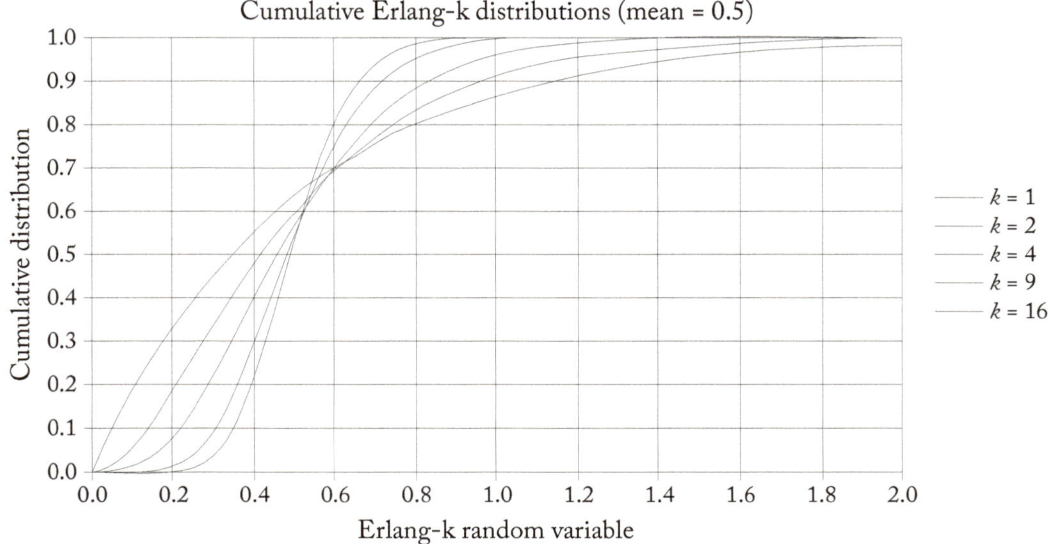

Figure 5.4: Various Erlang-*k* cumulative distributions.

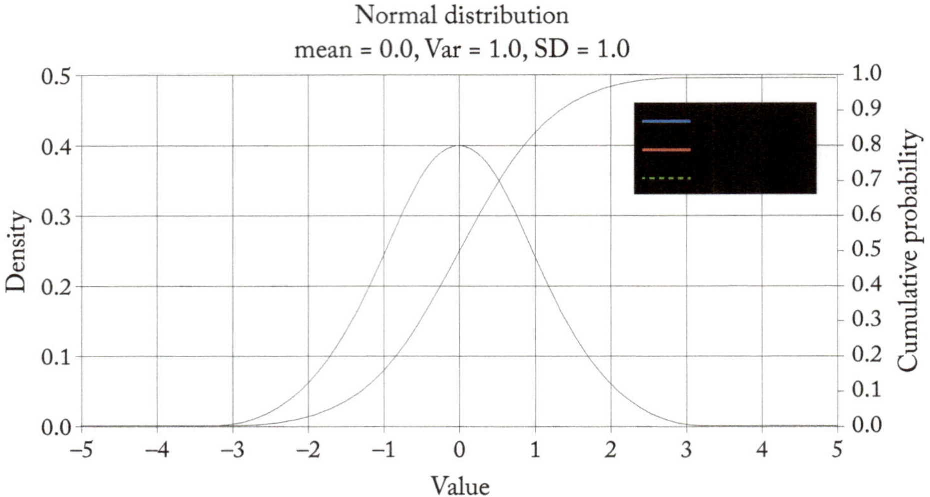

Figure 5.5: Standard normal density and cumulative functions.

5.5 THE NORMAL DISTRIBUTION

Finally, we turn to the *Normal* distribution. This is the (approximate) distribution of the sum of a large number of independent identically distributed random variables. The *standard Normal density* function is given by:

$$f_X(x) = \frac{1}{\sqrt{2\pi}} e^{-x^2/2} \quad -\infty < x < \infty.$$

This form of the Normal distribution has a mean or average of 0 and a standard deviation (a measure of the variability of the distribution) of 1. The following, more general form, has a mean of μ and a standard deviation of σ:

$$f_Y(y) = \frac{1}{\sqrt{2\pi}\sigma} e^{-(y-\mu)^2/(2\sigma^2)} \quad -\infty < y < \infty.$$

The cumulative Normal distribution must be computed numerically. Alternatively, there are tables of the cumulative standard Normal distribution. Any Normal distribution can be *standardized* by letting $x = \frac{y-\mu}{\sigma}$. We can then look up the cumulative distribution of x, which will be identical to the cumulative distribution of y.

Figure 5.5 plots the standard Normal density function (blue) and the corresponding cumulative distribution (red).

CHAPTER 6

Functions of a Random Variable

6.1 BASIC CONCEPTS AND NOTATION

In the preceding two chapters, we discussed discrete and continuous random variables, respectively. In particular, we introduced a number of often-used probability mass functions (for discrete random variables) and density functions (for continuous random variables). In this chapter, we explore the notion of *functions* of a random variable. This will allow us to formally introduce the concepts of the mean, variance, standard deviation, and coefficient of variation of a distribution. These are all ways of characterizing a random variable or process.

Suppose X is a random variable with either a probability mass function given by $p_X(x)$ or a density function given by $f_X(x)$. Now let $g_X(x)$ be a function of the random variable X. Since X is a random variable, the quantity $g_X(x)$ is also clearly a random variable. To illustrate this, suppose X is the number of phone messages you receive each day on your mobile phone. We might want to know the mean or average number of messages you receive as well as some measure of the variability of the number of messages you receive daily.

The expected value of a function of a random variable is the long-run average value of the function we would find if we performed the underlying experiment many times. Thus, on any given day, you clearly get some number of calls on your cell phone. The expected number is the average number of calls you get when we average over a very long period of time. More rigorously, the expected value of a function of a random variable is given by:

$$E[g_X(x)] = \sum_{\text{all } x} g_X(x) p_X(x)$$

if X is a discrete random variable and by

$$E[g_X(x)] = \int_{\text{all } x} g_X(x) f_X(x) dx$$

if X is a continuous random variable.

6.2 THE MEAN OF A RANDOM VARIABLE

The mean of a random variable is just the probability-weighted average value that the random variable can take on. To find the mean, we simply let $g_X(x) = x$ in the formulae above. Thus,

Table 6.1: Sample distribution of the number of daily messages received

Messages	Probability	Cumulative Probability	1-Cumulative Probability
0	0.15	0.15	0.85
1	0.25	0.40	0.60
2	0.27	0.67	0.33
3	0.18	0.85	0.15
4	0.10	0.95	0.05
5	0.05	1.00	0.00

we would get

$$E[X] = \sum_{\text{all } x} x p_X(x)$$

in the discrete case and

$$E[X] = \int_{\text{all } x} x f_X(x) dx$$

in the continuous case.

Let us illustrate this with a simple example. Suppose Table 6.1 gives the distribution of the number of phone messages you receive daily on your mobile phone. The expected number of calls you get is given by:

$$E[calls] = 0(0.15) + 1(0.25) + 2(0.27) + 3(0.18) + 4(0.10) + 5(0.05) = 1.98.$$

In other words, on average you receive slightly less than 2 calls per day, though there is a 15% chance that you will get 4 or more calls, or more than twice the average number.

We can also show that if X is a non-negative random variable, we can compute the mean using the following equations for discrete and continuous random variables, respectively:

$$E[X] = \sum_{x=0}^{\infty} \{1 - P_X(x)\} \qquad E[X] = \int_0^{\infty} \{1 - F_X(x)\} dx.$$

The reader should verify that the formula $E[X] = \sum_{x=0}^{\infty} \{1 - P_X(x)\}$ gives the mean for the probability mass function given in Table 6.1.

6.3 THE VARIANCE OF A RANDOM VARIABLE

The quantity $x - E[X]$ is a measure of how far a particular realization x is from the average or mean value of the random variable. Unfortunately, if we let $g_X(x) = x - E[X]$, the expected value of $g_X(x)$ will always be 0, since the positive and negative deviations from the mean will cancel each other out. But, if we square the deviations and take the expected value of the squared deviations, we will not get 0. Thus, we let $g_X(x) = (x - E[X])^2$. The expected value of this function is called the *variance* of a random variable. Thus, we have

$$Var(X) = \sum_{\text{all } x}(x - E[X])^2 p_X(x)$$

in the discrete case and

$$Var(X) = \int_{\text{all } x}(x - E[X])^2 f_X(x)dx$$

in the continuous case. These are not the easiest of formulae to navigate. Fortunately, there is an easier way of computing the variance. For any random variable, we can show that

$$Var(X) = E[X^2] - (E[X])^2,$$

where $E[X^2]$ is just given by $E[X^2] = \sum_{\text{all } x} x^2 p_X(x)$ in the discrete case and by $E[X^2] = \int_{\text{all } x} x^2 f_X(x)dx$ in the continuous case.

For the simple example shown in Table 6.1, we get

$$E[calls^2] = 0(0.15) + 1(0.25) + 4(0.27) + 9(0.18) + 16(0.10) + 25(0.05) = 5.8$$

and

$$Var(calls) = 5.8 - (1.98)^2 = 1.8796 \approx 1.88.$$

It is important to note at this point that unless the function $g_X(x)$ is linear, we will not in general have $E[g_X(x)] = g_X[E(X)]$. That is, in general it is not true that the expected value of a function is equal to the function evaluated at the mean of the random variable. In the case of the variance, $g_X(x) = (x - E[X])^2$ and $g_X[E(X)] = [E(X) - E(X)]^2 = 0$. But we know that the variance will always be strictly positive, unless we have a distribution that takes on only one value with probability 1. Thus, in general, we have

$$E[g_X(x)] \neq g_X[E(X)].$$

6.4 THE STANDARD DEVIATION OF A RANDOM VARIABLE

If you think carefully about the variance, you realize two things. First, the variance is always non-negative (and generally positive) since it is the expected value of a squared quantity, and squared quantities are always positive. Second, the units associated with the variance are given by the *square* of the units of the random variable. Thus, the variance of the number of calls you receive daily is measured in terms of *calls²* which is a very strange set of units. If we take the square root of the variance, we obtain the *standard deviation* of the distribution. The units of the standard deviation are the same as those of the random variable. The standard deviation of the number of calls you get in the example above is about 1.37 calls per day.

Table 6.2: Measures of income distributions

Metric	Sarah	Ann
Mean	$40,000	$5,000
Standard deviation	$5,000	$1,000
Coefficient of variation	0.125	0.20

6.5   THE COEFFICIENT OF VARIATION OF A RANDOM VARIABLE

Let us think about two individuals: Sarah is a stock broker and has a monthly income that is $34,000 with probability 0.5 and is $45,000 with probability 0.5 depending on whether or not she has a good month. Her average income is therefore $40,000. Ann's income is also uncertain. She is a nurse and either earns $4,000 per month or $6,000 per month, each with probability 0.5. Her income depends on how many overtime hours she works each month. Her average income is $5,000. Whose income is more variable?

Table 6.2 summarizes three key metrics for Sarah and Ann. Clearly, Sarah's income is, on average, 10 times Ann's income. The standard deviation of Sarah's income is five times that of Ann's income. Table 6.2 also includes the *coefficient of variation*, which is a dimensionless quantity. It is the ratio of the standard deviation divided by the mean. Based on this metric, Ann's income is more variable than is Sarah's. Another way of thinking about this is that the same amount of variability (variance or standard deviation) is more impactful when the mean is small than when it is very large.

6.6   METRICS FOR COMMON DISTRIBUTIONS

Table 6.3 summarizes the key discrete distributions we have discussed in Chapter 4 along with the mean, variance, and standard deviation of each of these distributions. Table 6.4 summarizes the key continuous distributions discussed in Chapter 5. Note that two forms are given for the Erlang-k distribution. Form 1 is simply the distribution of the sum of k independent identically distributed exponential random variables, each with parameter λ. Note that in this form, the mean grows linearly with the parameter k. In the next row we show the form outlined in Chapter 5. In this case, this is the sum of k independent identically distributed exponential random variables, each with parameter $k\lambda$. In this case, the mean is independent of the value of k.

Table 6.3: Summary of key discrete distributions

Distribution	Comment	Parameters	Probability Mass Function	Mean	Variance	Standard Deviation
Binomial	Number of successes in N trials	N, q	$\binom{N}{n} q^n (1-q)^{N-n}$ $n = 0,1,\ldots,N$	Nq	$Nq(1-q)$	$\sqrt{Nq(1-q)}$
Geometric	Number of failures until the first success	q	$(1-q)^n q$ $n = 0,1,\ldots.$	$\dfrac{1-q}{q}$	$\dfrac{1-q}{q^2}$	$\dfrac{\sqrt{1-q}}{q}$
Poisson	Limit of the binomial as $N \to \infty$ holding $Nq = \lambda$	λ	$\dfrac{\lambda^n e^{-\lambda}}{n!}$ $n = 0,1,\ldots$	λ	λ	$\sqrt{\lambda}$

Table 6.4: Summary of key continuous distributions

Distribution	Comment	Parameters	Density Function	Mean	Variance	Standard Deviation
Continuous uniform	Equally likely outcomes between A and B	A, B	$\dfrac{1}{B-A}$ for $A \le x \le B$ 0 elsewhere	$\dfrac{A+B}{2}$	$\dfrac{(B-A)^2}{12}$	$\dfrac{B-A}{2\sqrt{3}}$
Exponential	Key memoryless distribution	λ	$\lambda e^{-\lambda t}$ for $t \ge 0$	$\dfrac{1}{\lambda}$	$\dfrac{1}{\lambda^2}$	$\dfrac{1}{\lambda}$
Erlang-k (form 1)	Sum of k independent identically distributed exponential random variables with parameter λ	k, λ	$\dfrac{\lambda(\lambda t)^{k-1} e^{-\lambda t}}{(k-1)!}$ for $t \ge 0$	$\dfrac{k}{\lambda}$	$\dfrac{k}{\lambda^2}$	$\dfrac{\sqrt{k}}{\lambda}$
Erlang-k (form 2)	Sum of k independent identically distributed exponential random variables with parameter $k\lambda$	$k, k\lambda$	$\dfrac{k\lambda(k\lambda t)^{k-1} e^{-k\lambda t}}{(k-1)!}$ for $t \ge 0$	$\dfrac{1}{\lambda}$	$\dfrac{1}{k\lambda^2}$	$\dfrac{1}{\sqrt{k}\lambda}$
Normal	Sum of a large number of independent random variables	μ, σ	$\dfrac{1}{\sqrt{2\pi}\sigma} e^{-(y-\mu)^2/(2\sigma^2)}$ $-\infty < y < \infty$	μ	σ^2	σ

C H A P T E R   7

Fundamentals of Optimization

7.1   OPTIMIZATION BASICS

As indicated in Chapter 1, one of the fundamental building blocks of operations management is optimization, or at least an effort to improve upon the current situation. Throughout much of the remainder of this text, we will be using optimization modeling. This chapter introduces the basics of optimization. Before you get overly concerned about this, you have probably seen optimization before, though it may not have been called that. As we will see, when you take a derivative of a function to find a (local) minimum or maximum, you are optimizing the function (unless you hit a saddle point of the function). Later in this chapter, we will review the basics of linear regression, which many of you may have seen as well. In our case, we will emphasize that we are optimizing a function when we solve for the linear regression coefficients. In any event, don't panic. Optimization is not all that hard, at least from our perspective.

At least half of the battle in solving any optimization problem is identifying four key elements of the model.

(a) The **inputs** are the quantities that you know going into the problem. For example, if you are trying to figure out how many ambulance bases to have in your local city, a key input would be the population of each Census block group. In Ann Arbor, MI, there are 98 block groups with population values ranging from 110 to 2,924, with an average of about 1,231. You are also likely to know the distances or travel times between the centroids of the block groups. You are likely to have some service standard as well. For example, you might want everyone (or at least the centroid of very block group) to be within 5 minutes of the nearest ambulance base. Finally, you might also have the cost of building an ambulance base in each block group.

(b) The **decision variables** are the quantities you need the model to determine for you. When locating ambulances in a city, there might be a decision variable for each block group indicating whether or not we locate an ambulance in that block group.

(c) The **objective function** is a function of the decision variables (and perhaps some of the inputs) and represents what you are trying to achieve. For example, we may want to minimize the number or cost of the ambulances needed to ensure that everyone is within 5 minutes of the nearest ambulance base.

(d) The **constraints** represent conditions that inhibit or restrict our ability to achieve the objective function. For example, the minimum number of ambulances needed would be 0 until we add a constraint that says that every block group has to be within 5 minutes of the nearest ambulance base. This constraint forces us to locate enough ambulance bases to satisfy this condition.

Identifying these key components of an optimization problem—the *inputs* and *decision variables* on the one hand and the *objective function* and the *constraints* on the other hand—is, in many cases, the hardest part of solving an optimization problem. Once this is done, there are often some excellent software packages that perform the numerical work for us. Some are even built into Excel, though the Solver in Excel is limited in the size of the problems it can solve and in its overall capabilities. There are much more powerful add-ins to Excel and stand-alone packages as well as routines that can be called from languages like Python.

In the remainder of this book, we will see numerous optimization examples. For now, we turn our attention to a problem in statistics, finding the best line to fit a set of observations. In statistics, this is called *regression*.

7.2   REGRESSION AS OPTIMIZATION

Consider the problem faced by the campaign manager for a major presidential candidate. In the United States, presidential elections are not won based on the popular vote, but rather based on the Electoral College. There are 538 electoral votes up for grabs each Presidential election and a candidate needs 270 to win the election. The number of electoral votes allotted to each state is equal to the number of representatives the state has in the House of Representatives plus two for each of the two senators that each state has. There are about 760,000 people per representative. In addition, Washington, DC, with a population of about 690,000 people but no official representation in either the House of Representatives or the Senate, gets 3 electoral votes. In all but two states, Maine and New Hampshire, all of the electoral votes of a state go to the candidate who receives the most votes in the state. Thus, it is possible for a candidate to win the popular vote, but to lose the election in the Electoral College, and therefore not be elected President of the United States.

To see how this can happen, consider just three states: Alabama, Minnesota, and Nevada, as shown in Table 7.1. In this hypothetical "country," there are only 25 electoral votes up for grabs. Clearly, a candidate must win at least any two of the three states to win the election. In this hypothetical set of results (which are close to the vote totals for the three states in the 2020 election), the Republican candidate won the popular vote, but the Democratic candidate won the election since s/he garnered 16 of the 25 Electoral Votes. This sort of result is *not* just theoretically possible. It has happened five times in U.S. history, most recently in the 2000 and 2016 elections when the Republican candidate (George W. Bush in 2000 and Donald Trump in 2016) lost the popular vote but won the Presidency in the Electoral College.

Table 7.1: Sample hypothetical election results for three states

	Electoral Votes	Republican Votes	Democratic Votes	Republican EV	Democratic EV	Percent Democratic
Alabama	9	1,400,000	850,000	9	0	37.8%
Minnesota	10	1,480,000	1,720,000	0	10	53.8%
Nevada	6	670,000	700,000	0	6	51.1%
Total	**25**	**3,550,000**	**3,270,000**	**9**	**16**	

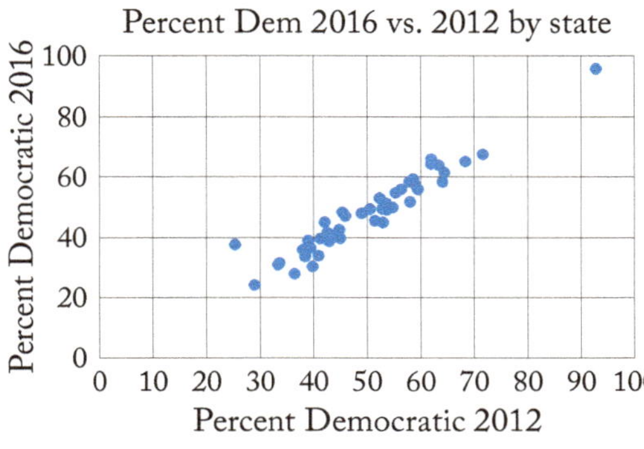

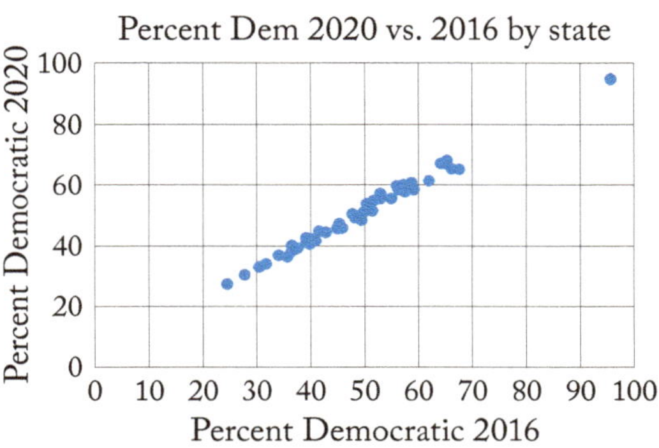

Figure 7.1: 2012 vs. 2016 percent Democratic and 2016 vs. 2020 percent Democratic by state.

If there were no relationship between how a state voted in one presidential cycle and the way the same state voted in the next presidential election four years later, the candidates in our hypothetical country would need to devote about equal energy to each of the three states. If there is a strong relationship between how a state voted in year X and how the state votes in year $X + 4$, then the Democrats can effectively forget about Alabama since it will almost surely end up casting its 9 Electoral Votes for the Republican candidate. The "battleground" states then become Minnesota and Nevada. The Republican candidate would have something of an advantage because he or she could cede one of the two states and simply devote all of his or her funding and campaign appearances to the other state. In fact, if the Republican candidate could just convince a little over 15,000 Democratic voters in Nevada to vote Republican, the Republican candidate would win the election. The Democratic candidate, on the other hand, must fight for votes in both Minnesota and Nevada if he or she wants to win the election.

Thus, the key question is how well can we predict the way a state will vote in year $X + 4$ based on how the state voted in year X. Figure 7.1 shows two plots. On the left we plot the percent of the votes that were Democratic in 2012 (X axis) vs. the percent of the votes that were Democratic in 2016 for each of the 50 states plus Washington, DC. The right plot gives the

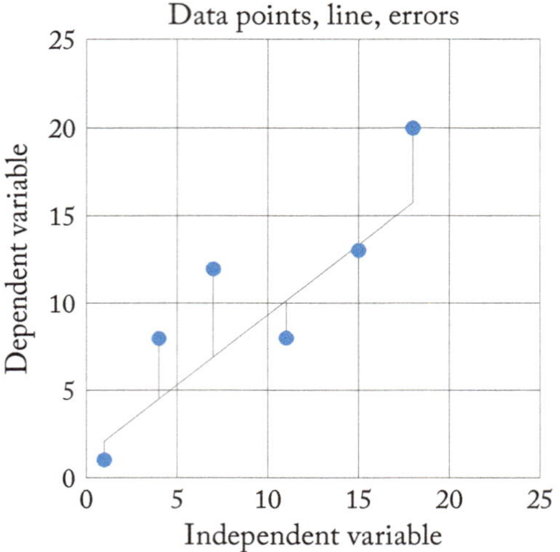

Figure 7.2: Data points, sample line, and sample errors.

same information for the 2016 and 2020 election years. The two plots clearly show that there is a strong relationship between how a state votes in year X and how the state will vote in year $X + 4$. This is not at all surprising.

We now want to find out if we can predict how a state will vote in a future presidential election based on how it voted in the previous election. Before doing so, we will look at a more generic case of regression. In Figure 7.2 we plot six data points. The variable on the X axis is called the *independent variable*. This is what we will generally know. The variable on the Y axis is called the *dependent variable*. We will have some observations for the dependent variable corresponding to specific values of the independent variable. What we would like to do is to identify a relationship between the dependent and independent variables, so that we can then predict the value of the dependent variable given a new observation or value for the independent variable. For example, we might want to look at the relationship between a person's height (the independent variable) and the individual's weight (the dependent variable). It is easy to observe a person's height. If we know the relationship between height and weight, we can then make an educated guess about the individual's weight.

In Figure 7.2, we see six observations or pairs of values for the independent and dependent variables. We also see a hypothetical line through the data. The question is this: is this the "best" line and how would we define "best"? To begin to define what the best line is, we can look at the errors or the difference between the observed value, y_j, of the jth data point and the predicted value $\hat{y}_j$ on the line. But the equation of the line is given by $\hat{y}_j = A + Bx_j$, where A and B are the unknown intercept and slope of the line. (The line in Figure 7.2 is given by $\hat{y}_j = 1.3 + 0.8x_j$.) In other words, we can look at $y_j - \hat{y}_j$ or $y_j - (A + Bx_j)$. Note, however,

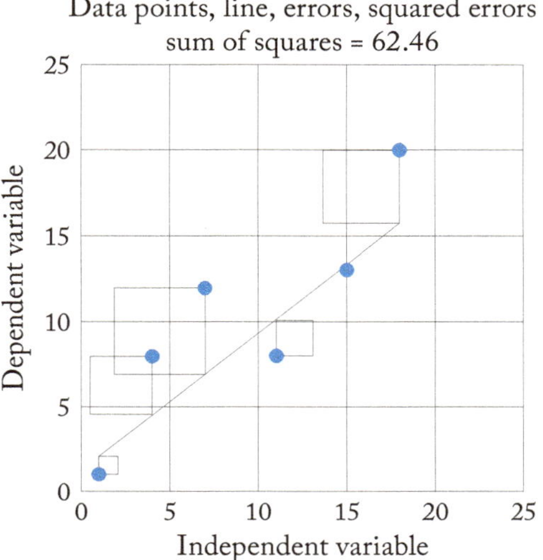

Figure 7.3: Sample data points, sample line, and "squared" errors.

that if we set $A = \bar{y}$, the average y value, and $B = 0$, the sum of the errors will always be 0. Therefore, we generally look at the sum of the *squared* errors. (This is analogous to squaring the difference between a value and the mean value in computing the variance as discussed in Chapter 6.) In other words, we want to find the values of A and B to minimize

$$\text{Sum of Squared Errors} = SS(A, B) = \sum_{j=1}^{n} \{y_j - (A + Bx_j)\}^2.$$

Here n is just the number of observations or six in our case.

Figure 7.3 plots the same six observations shown in Figure 7.2 and the same line, but now we are showing a square along with each data point. The sum of the areas of these squares is the value of the Sum of Squared Errors in the equation above. Erkut and Ingolfsson [2000] proposed this approach to visualizing the sum of squares in regression.

But the Sum of the Squared Errors is just a quadratic function in the unknowns A and B. To find the optimal values, we take the partial derivatives of SS with respect to A and B, set them to 0:

$$\frac{\partial SS(A, B)}{\partial A} = \sum_{j=1}^{n} 2\{y_j - (A + Bx_j)\} = 2\left\{\sum_{j=1}^{n} y_j - nA - B \sum_{j=1}^{n} x_j\right\} = 0$$

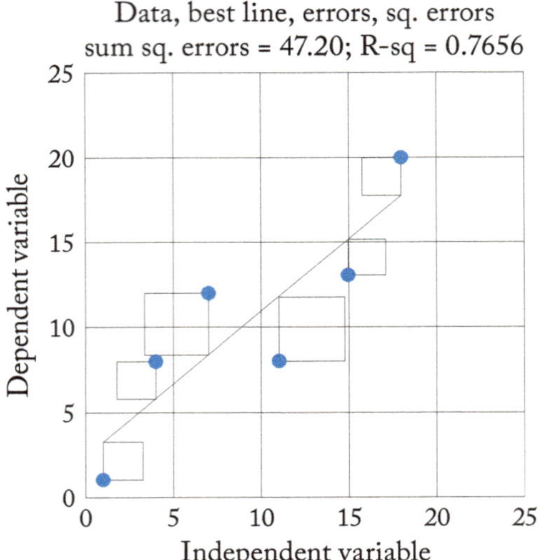

Figure 7.4: Sample data, best line, and squared errors.

$$\frac{d\,SS(A, B)}{\partial B} = \sum_{j=1}^{n} 2x_j \{y_j - (A + Bx_j)\} = 2\left\{\sum_{j=1}^{n} x_j y_j - A \sum_{j=1}^{n} x_j - B \sum_{j=1}^{n} x_j^2\right\} = 0.$$

When we solve these equations for A and B, we obtain:

$$A = \bar{y} - B \cdot \bar{x}$$

$$B = \frac{\displaystyle\sum_{j=1}^{n} x_j y_j - n \cdot \bar{x} \cdot \bar{y}}{\displaystyle\sum_{j=1}^{n} x_j^2 - n \cdot \bar{x}^2}.$$

Again, $\bar{x}$ and $\bar{y}$ are the average of the x and y values, respectively.

For the example in Figure 7.2, we obtain, $\hat{y}_j = 2.4 + 0.85x_j$. Figure 7.4 shows this line, the sample data points, and the squared errors. The sum of the squared errors is 47.2, which is considerably better than the value associated with the line shown in Figure 7.3. The R^2 value is a measure of how well the line fits the data. It is always between 0 and 1 with larger values indicating a better fit. It represents the fraction of the (squared) variation in the observations that is captured by the line.

Now that we know how to find the regression lines, let us figure out what they are for the two plots in Figure 7.1. Figures 7.5 and 7.6 plot the results. States that were Democratic in

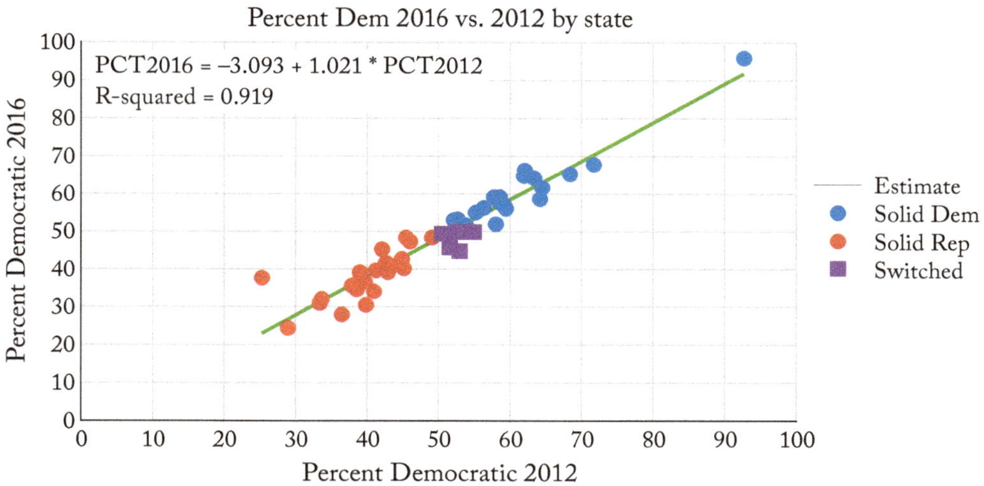

Figure 7.5: Regression of percent Democratic in 2016 vs. percent Democratic in 2012 by state.

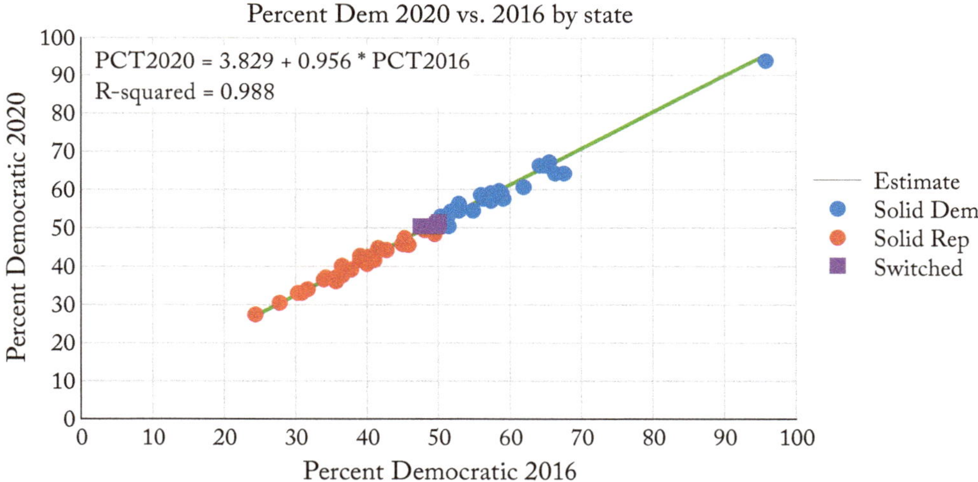

Figure 7.6: Regression of percent Democratic in 2020 vs. percent Democratic in 2016 by state.

year X and Democratic in year $X + 4$ are plotted in blue; states that were Republican in both years are plotted in red; states that switched are plotted in purple as squares. The very high R^2 values indicate that we have a very strong fit. The lines also suggest that the percent Democratic in a state in year $X + 4$ is virtually identical to the percent Democratic in the state in year X, **shifted by a constant**. In 2016, that constant was -3.093 indicating (roughly speaking) that

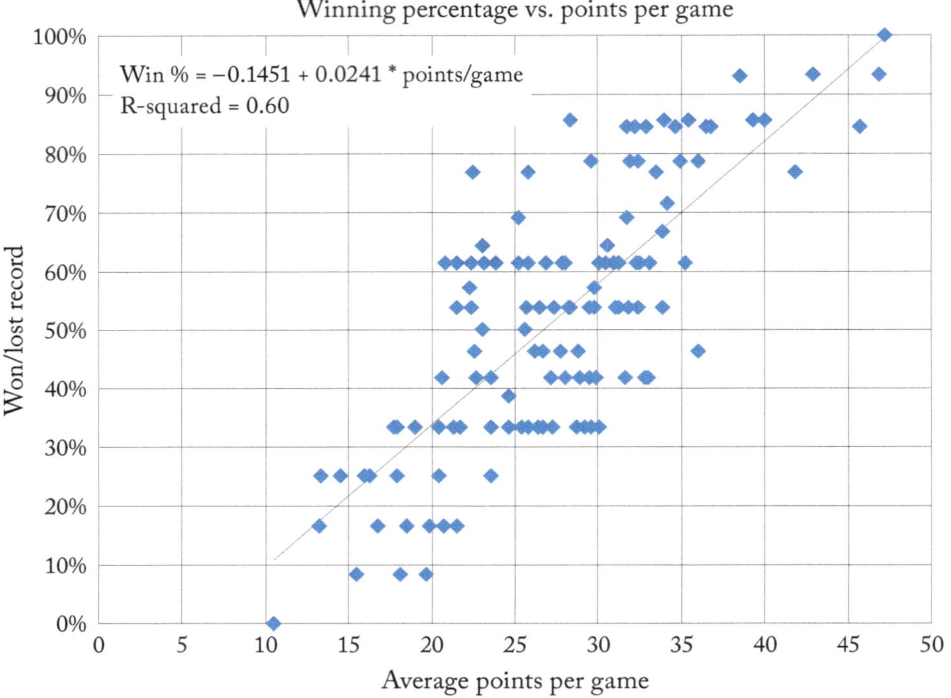

Figure 7.7: 2019–2020 NCAA football season won/loss record vs. points per game.

states that were over 53% Democratic in 2012 were (almost surely) going to vote Democratic in 2016. In 2020, the constant was 3.829, indicating that states that were under 46% Democratic in 2016 were likely to remain Republican in 2020. Loosely speaking, these results suggest that the battleground states are likely to be those with winning percentages of about 45–55%. In fact, this is what we often see. Candidates focus their attention on such states. In the 2020 election, there were campaign events in only 17 of the 50 states. There were no events in California and New York, the first and fourth most populous states. In 2016 they voted 66% and 62% Democratic. Why would either candidate waste time in such states? Pennsylvania and Florida— states the Democrats barely lost in 2016—saw the most campaign events with 47 and 31 events, respectively.[1] The Democratic candidate, Joe Biden, took Pennsylvania, but failed to flip Florida to the Democratic column in 2020.

As a final example of regression, consider the data shown in Figure 7.7. Here we plot the winning percentage of each of 130 NCAA Division 1 football teams vs. the average number of points the team scored per game. The best regression line is shown on the figure along with the

---

[1] https://www.nationalpopularvote.com/map-general-election-campaign-events-and-tv-ad-spending-2020-presidential-candidates. Accessed 5/29/2021.

R^2 value of 0.6. While this is a respectable R^2 value, the fact that it is not closer to 1 suggests that there may be other factors at play in determining a team's won/loss record. Such factors include measures of the team's defensive capability and the performance of their special teams.

7.3   REFERENCES

E. Erkut and A. Ingolfsson, Let's put the squares in least-squares, *INFORMS Transactions on Education*, 1(1):47–50, 2000. DOI: 10.1287/ited.1.1.47. 39

CHAPTER 8

Basic Inventory Management

8.1 FUNDAMENTALS OF INVENTORY MANAGEMENT

One of the key issues addressed by operations management is that of how a company should manage its inventory. From our perspective, this is a useful and gentle way of easing into optimization modeling in operations management.

First, why should a company hold any inventory? After all, holding inventory ties of valuable capital that could be used for other endeavors by the company. Companies must pay insurance costs for inventory they hold; the more valuable the inventory, the higher the insurance costs. Holding inventory also ties up valuable space in warehouses or fulfillment centers. Inventory may also be perishable. Clearly fresh fruits, vegetables, meat, fish, poultry, and other foods are perishable. But, other products are also perishable. Most drugs, for example, are perishable and must be discarded after an expiration date. An employee at a camera store in suburban Chicago once told me that by the time they received their first shipment of some Japanese cameras, the company had already announced an upgraded model. At the end of a model year, auto dealers often sell last year's model at a discount to make room for the new models. Even durable goods are perishable.

So why do companies hold inventory if there are all these costs associated with doing so? From our perspective, the answer boils down to two issues: *tradeoffs* and *uncertainty*. To begin, there are some items that are not sold individually. Generally speaking, you cannot go into a grocery store and buy a single plastic fork; you have to buy a box of plastic forks, which may contain 24 or more forks. Thus, you are forced to maintain an inventory of plastic forks at your home. Second, there are costs associated with placing and receiving orders for goods. Imagine how large the shipping costs would be to your local coffee shop if they bought coffee in one-pound packages one at a time. Each one-pound package would incur shipping and handling costs. Surely it would be cheaper for them to order coffee in larger quantities, perhaps 20-pound bags, purchasing many such bags at a time spreading the shipping and handling costs over a much larger quantity. Also, there are administrative costs associated with placing orders from vendors. Surely, your local coffee shop would not want an employee devoted full-time to placing orders for coffee every time a new pound of coffee was used (which could be 20 or more times per hour in a busy coffee shop). In short, there are many fixed costs associated with inventory management. Part of the role of operations management is to find the appropriate *tradeoff* between the inventory holding costs and the fixed costs of placing orders for the inventory.

Uncertainty is another reason to hold inventory. Your local coffee shop does not know *a priori* what the demand for coffee will be on any given day. To avoid running out of coffee, they maintain an inventory of coffee, cups, lids, sugar, and cream. At a much larger scale, the demand for the COVID vaccine was uncertain when the federal government contracted with pharmaceutical companies to produce the vaccine. Major hospitals, like the University of Michigan hospital system, use over 3,000 different drugs every day. The daily demand for any one drug is uncertain, but the cost of running out of a drug is very large. Therefore, hospitals maintain an inventory of drugs to protect against shortages. This must be balanced (*traded off*) against the costs associated with wasting expired drugs [Czerniak et al., 2021]. Not only is demand often uncertain, but supply can also be uncertain. In the case of the COVID vaccine, the number of successful efforts to develop the vaccine was uncertain when the pandemic began and when the government contracted with companies to produce the vaccine. Production problems have added to the vaccine supply uncertainty. Modern automobiles use many computer chips per vehicle. As I write this, there is a global shortage of such chips and most auto manufacturers have cut back on production, often leading to idled plants. Thus, *uncertainty* is another reason to hold inventory [Boston, 2021].

In the remainder of this chapter, we examine a very basic inventory management model that accounts for the key tradeoffs outlined above, but that ignores uncertainty. In the next chapter, we will incorporate demand uncertainty, again in a classic inventory model.

8.2   THE ECONOMIC ORDER QUANTITY MODEL

The *economic order quantity* (EOQ) model is the most basic of all inventory models. In this model, we make the following assumptions.

- The **lead-time** is 0, meaning that as soon as an order is placed, the order arrives. This is an easy assumption to relax.

- The demand is **deterministic**, meaning that there is no uncertainty in the demand.

- The demand is **static**, meaning that the demand is constant and does not change over time.

- There are no **backorders**, meaning that all demand is satisfied from current inventory.

- We are dealing with a **single SKU** (or stock-keeping-unit). In other words, placing orders for dark-roast coffee will be handled separately from ordering medium-roast coffee or decaf-coffee.

- There are no **quantity discounts**, meaning that if we order a single pound of coffee or 10,000 pounds of coffee in a single order, the unit cost per pound is the same.

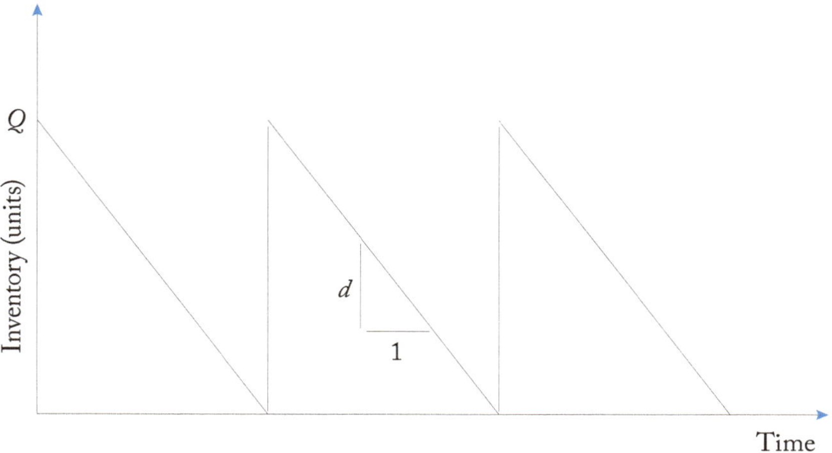

Figure 8.1: Inventory over time in the EOQ Model.

- Production is **instantaneous**, meaning that once an order is placed, the entire quantity arrives in one batch and, by the first assumption about the lead time, the inventory arrives as soon as the order is placed.

With these assumptions, we introduce the following notation:

INPUTS

d demand rate (in units per year)
c production cost (not counting any fixed setup costs) in \$/unit
f fixed setup cost in \$/order
h holding cost per unit per year

DECISION VARIABLES

Q purchase size or order size or order quantity

Figure 8.1 plots the inventory level as a function of time using this notation. When an order is placed, we instantly receive Q items. This is depleted at a rate of d items per year. When the inventory hits 0, a new order is placed and, according to the assumptions above, the order arrives instantly and the inventory level immediately goes up to Q.

Each year, we must place d/Q orders. If we use 12,000 paper cups each year at a coffee shop and we order 500 cups in each order, we would need to place 24 orders each year. Thus, the annual cost of placing orders is given by $f\frac{d}{Q}$, meaning that the ordering cost goes down with the order size, Q.

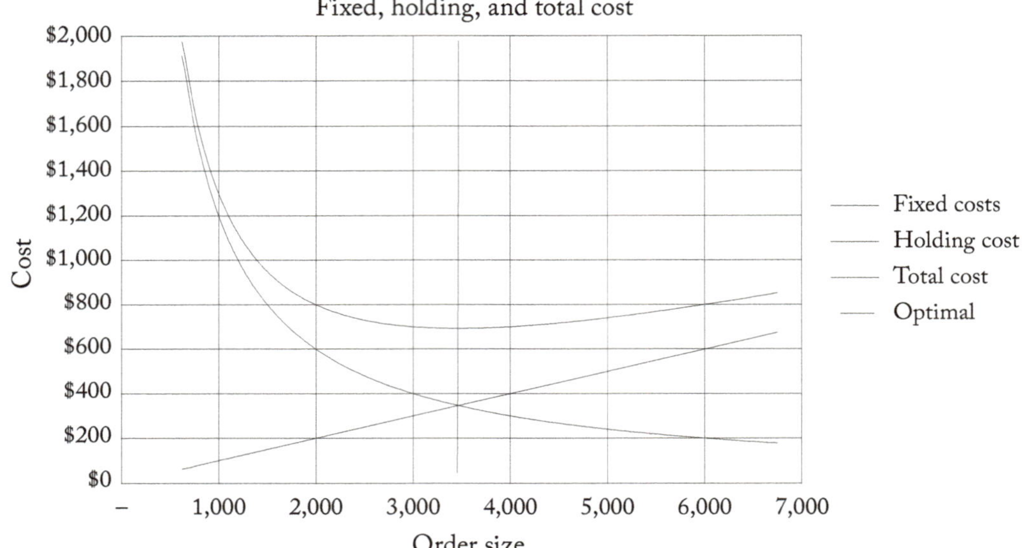

Figure 8.2: Example fixed, holding, and total cost.

Looking at Figure 8.1, the average inventory on hand is $Q/2$. Each unit of inventory costs h per year. Therefore, the annual inventory holding cost is given by $h\frac{Q}{2}$. Suppose the annual demand for a product is 12,000 units, the fixed cost of placing and order is \$100 per order, and the holding cost per item per year is \$0.20. Figure 8.2 plots the fixed ordering costs (blue), holding costs (red), and total cost (green) as a function of the order size. Initially, as the order size increases, the fixed cost decreases rapidly. After a certain point, the decrease in fixed cost becomes more modest. The holding cost increases linearly with the order size. The total cost curve (green) is U-shaped.

The total cost—the sum of the ordering cost and the holding cost—is given by:

$$TC(Q) = f\frac{d}{Q} + h\frac{Q}{2}.$$

To find the optimal order quantity, we simply take the derivative of this function with respect to the order quantity, Q, and equate the result to 0. This gives,

$$\frac{dTC(Q)}{dQ} = -\frac{fd}{Q^2} + \frac{h}{2} = 0.$$

Solving for the optimal order quantity, we get

$$Q^* = \sqrt{\frac{2fd}{h}}.$$

So, the optimal order size increases with the cost of placing an order (f) and with the annual demand (d) and decreases with the unit holding cost (h), as expected. Substituting this optimal order quantity into the total cost function, we obtain

$$TC(Q^*) = fd\sqrt{\frac{h}{2fd}} + \frac{h}{2}\sqrt{\frac{2fd}{h}}$$

$$= \underbrace{\sqrt{\frac{fdh}{2}}}_{\text{from order cost}} + \underbrace{\sqrt{\frac{fdh}{2}}}_{\text{from holding cost}}$$

$$= \sqrt{2fdh}.$$

Note that at the optimal order quantity, the fixed order costs and the holding costs are equal in this model, as shown in Figure 8.2.

Also note, when we look at tradeoffs, it will not generally be true that the optimum occurs when the costs being traded off are equal. We will generally have to use the approach outlined above which involves (at a minimum) taking the derivative of the total cost function and equating it to 0.

8.3 SENSITIVITY OF THE ECONOMIC ORDER QUANTITY MODEL TO CHANGES IN THE ORDER QUANTITY

For the example of Section 8.2, the optimal order size is 3,464 and the optimal cost is $692.82 per year. These are clearly not very convenient numbers. We might want to know how much worse the total cost would be if we ordered 3,000 units (four times per year) or 4,000 (three times per year). More generally, if we order αQ^* instead of Q^*—in other words, if we order a multiple α of the optimal order quantity—we can show that

$$\frac{TC(\alpha Q^*)}{TC(Q^*)} = \left(\frac{\alpha}{2} + \frac{1}{2\alpha}\right).$$

Figure 8.3 plots this ratio. For a wide range of values of α the ratio is very close to 1.0, meaning that the penalty for ordering in quantities that differ from Q^* is relatively low. In particular, in the example above, if the order size is 3,000, the total cost goes up to $700, a 1% increase. If we order 4,000 units at a time, the total cost again goes up to $700. Figure 8.4 is identical to Figure 8.2 except that we are now showing the region in which the total cost is within 2.5% of the optimal total cost. This is clearly a large region. Given the numerous assumptions made in the EOQ model, using an order size or quantity that is within 2.5% of the optimal total cost, for reasons of convenience, is acceptable in just about any set of circumstances.

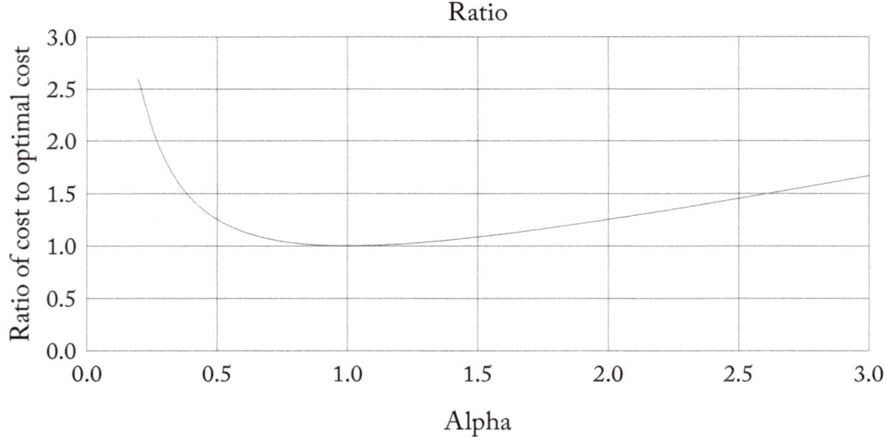

Figure 8.3: Ratio of total cost to optimal cost if the order size is alpha times the optimal order size.

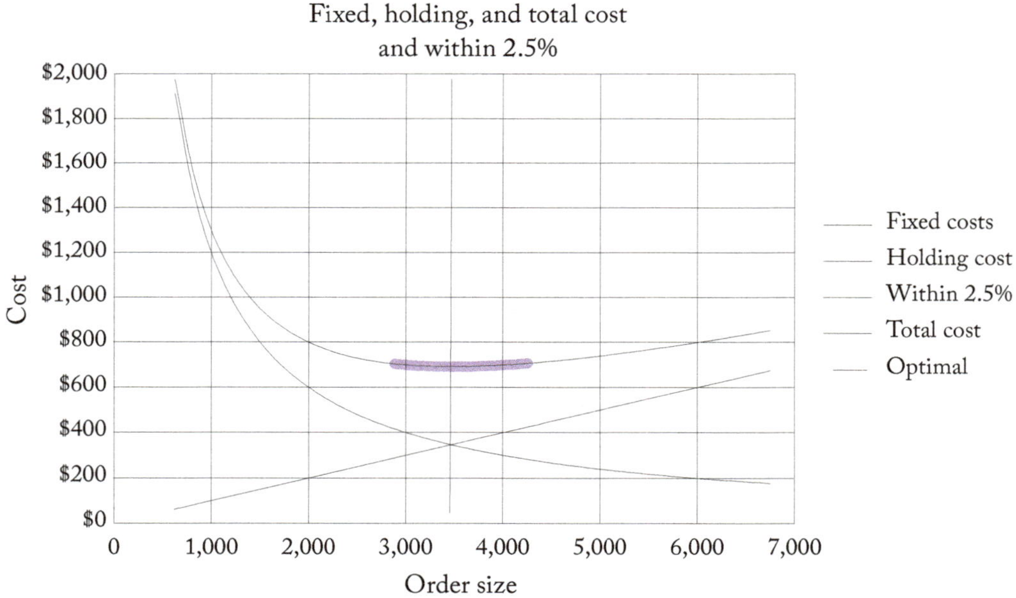

Figure 8.4: Example cost components and region in which total cost is within 2.5% of the optimal cost.

8.4   EXTENSIONS OF THE EOQ MODEL

The EOQ model makes many simplifying assumptions as outlined in Section 8.2. All of these assumptions can be relaxed relatively easily, though doing so is generally beyond the scope of this text. Relaxing some of these assumptions (including the backordering assumption) are simple extensions of the model outlined above. Other assumptions require alternative modeling approaches.

One easy assumption to relax is the assumption that the lead-time is 0. If the lead-time is equal to τ days, and if the lead time is deterministic or non-random, then we should simply place an order when the inventory level reaches $\frac{\tau d}{365}$. In this way, the order will arrive exactly when the inventory level reaches 0, as in the standard EOQ model.

8.5   REFERENCES

W. Boston, Global chip shortage set to worsen for car makers, *Wall Street Journal*, April 29, 2021. 46

L. L. Czerniak, M. S. Daskin, M. S. Lavieri, B. V. Sweet, J. Erley, M. A. Tupps, Improving simulation optimization run time when solving for periodic review inventory policies in a pharmacy, *Proc. of the Winter Simulation Conference*, S. Kim, B. Feng, K. Smith, S. Masoud, Z. Zheng, C. Szabo, and M. Loper, Eds., 2021. 46

CHAPTER 9

Inventory Management with Uncertain Demand

9.1 ONE TAKE ON MODELING DEMAND UNCERTAINTY

In Chapter 8, we introduced inventory theory. We noted that there is often a tradeoff between the fixed costs of placing an order (which will drive us to having fewer orders of larger size) and the inventory holding costs (which will drive us to more frequent orders of smaller size). This *tradeoff* led to the *economic order quantity* model. This model, however, like many inventory models, assumes that the demand is deterministic. That is, the EOQ model assumes there is no uncertainty in the demand.

In this chapter, we introduce one way of relaxing the assumption that demand is deterministic. We will assume that we have only one chance to place an order with the supplier. While this may seem like a heroic assumption, in fact, this occurs in many situations. The classic example, from which the model gets its name, is that of a local newsstand or convenience store placing orders for newspapers from the publisher. Typically, they place a single order for some number of newspapers. They buy each paper at a cost of c, and sell newspapers at a price of p, with $p > c$, meaning that they make a profit of $p - c$ on each paper they sell. However, demand is uncertain, though there is a known probability distribution associated with the demand. If the store orders too few newspapers, there will be sales that are lost and profit that is foregone. On the other hand, if the store orders too many papers and the demand is less than the number of papers they buy from the publisher, any unsold papers can be returned to the publisher for a salvage value of s, where $s < c$. In other words, they lose $c - s$ on each unsold paper.

Most of us will not spend our lives selling newspapers, and almost certainly not in this manner. However, many other items operate this way. For example, seasonal, high-end, fashion goods are often ordered from manufacturers months in advance of the sales season. Retailers have only one chance to order the goods in appropriate styles, colors, and sizes. They then receive the orders and begin selling the clothing. Again, demand is uncertain and they will make a profit of $p - c$ on each item sold, but will lose $c - s$ on each unsold item at the end of the season. These items may be given to a thrift shop or sold at a significant discount.

Similarly, many employers allow employees to put funds aside for qualified medical expenses. Some time in the fall of year $T - 1$, employees must elect how much to set aside during the coming year, year T. The advantage of doing so is that money put into such a fund is tax exempt. Thus, every dollar spent from the fund is a pre-tax dollar. However, your healthcare expenses during year T are unknown when you decide how much to set aside. Loosely speaking, any unused funds at the end of the year are lost. (Actually, you can often spend those funds through mid-March of year $T + 1$, or in some cases, you may be able to roll over a part of any unused funds to year $T + 1$.) This again is a form of newsvendor problem.

Finally, college admission decisions are, in some sense, like a newsvendor problem. Admission officers are tasked with filling an incoming class. If too few students matriculate (actually enroll), tuition dollars are lost. If too many enroll, there are often unplanned costs for housing the extra students and creating class sections for the students. The yield (or percent of admitted students who enroll) is not known *a priori* at the time admission decisions are made. Even if the percent yield is known (e.g., if 40% of the admitted students matriculate), a college that admits 10,000 students hoping to fill a class of 4,000 would find that in 90% of the years, the number of matriculating students would be between 3,896 and 4,103. This is quite a large range, and there is a 10% chance that the number of matriculating students would fall outside this range. As in many other cases in which the newsvendor problem arises, college admission officers have a large number of tools available to them to reduce this uncertainty, including: early admission offers and wait lists.

9.2   THE NEWSVENDOR PROBLEM

Having established that the newsvendor problem arises in may practical instances, we now turn to the mathematics behind the newsvendor problem. Again, we define the following inputs:

INPUTS

p	unit sale price
c	unit cost
s	unit salvage value
D	a random variable representing the demand
$q(d)$	probability that the demand is exactly d units
$Q(d)$	probability that the demand is less than or equal to d units

The single decision variable is defined as follows:

DECISION VARIABLE

B	number to buy

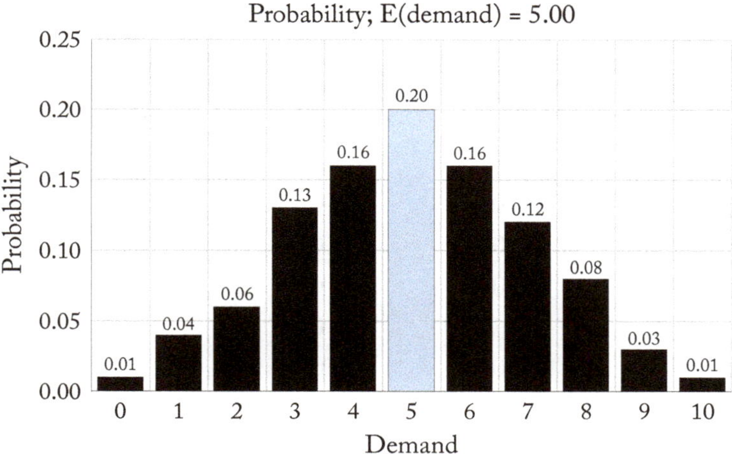

Figure 9.1: Sample probability mass function of demand.

Our goal is to maximize the expected profit or revenue minus the cost. This is given by:

$$\Pi(B) = -cB + p \sum_{d=0}^{B} d \cdot q(d) + pB\{1 - Q(B)\} + s \sum_{d=0}^{B} (B - d) \cdot q(d)$$

$$= \underbrace{-cB}_{\substack{\text{cost of}\\\text{purchased}\\\text{items}}} + p \underbrace{\sum_{d=0}^{\infty} \min(d, B) \cdot q(d)}_{\text{revenue from sales}} + s \underbrace{\sum_{d=0}^{B} (B - d) \cdot q(d)}_{\text{revenue from salvaged items}}.$$

The first term, $-cB$, represents the cost of buying B items from the supplier or manufacturer. The second term in the second line (or the second and third terms in the first line), $p \sum_{d=0}^{\infty} \min(d, B) \cdot q(d)$, represents the revenue from items that are sold. The last term, $s \sum_{d=0}^{B} (B - d) \cdot q(d)$, represents the revenue from salvaged items.

We will illustrate this with the following example. Suppose we have $p = 100$, $c = 40$, and $s = 25$. The demand can go from 0–10 items and the probability mass function is given in Figure 9.1. The expected demand is exactly 5. Table 9.1 shows the necessary calculations if we buy $B = 5$ items. This corresponds to buying an amount that equals the expected demand. When we do this, the expected profit is

$$\Pi(5) = -40 \cdot 5 + 100\{0 \cdot (0.01) + 1 \cdot (0.04) + \cdots + 5 \cdot (0.2) + 5 \cdot (0.4)\}$$

$$+ 25\{5 \cdot (0.01) + 4 \cdot (0.04) + \cdots + 1 \cdot (0.16)\}$$

$$= -200 + 419 + 20.25 = 239.25.$$

Table 9.1: Computation for $B = 5$ and sample probability mass function

Demand	Probability	Cumulative	Number Sold	Number Salvaged
0	0.01	0.01	0	5
1	0.04	0.05	1	4
2	0.06	0.11	2	3
3	0.13	0.24	3	2
4	0.16	0.40	4	1
5	0.2	0.60	5	0
6	0.16	0.76	5	0
7	0.12	0.88	5	0
8	0.08	0.96	5	0
9	0.03	0.99	5	0
10	0.01	1.00	5	0

Table 9.2: Computation for $B = 7$ and sample probability mass function

Demand	Probability	Cumulative	Number Sold	Number Salvaged
0	0.01	0.01	0	7
1	0.04	0.05	1	6
2	0.06	0.11	2	5
3	0.13	0.24	3	4
4	0.16	0.40	4	3
5	0.20	0.60	5	2
6	0.16	0.76	6	1
7	0.12	0.88	7	0
8	0.08	0.96	7	0
9	0.03	0.99	7	0
10	0.01	1.00	7	0

Is this the best we can do? Actually, it is not. We should order seven items with the following expected profit. Table 9.2 shows the key computations in this case.

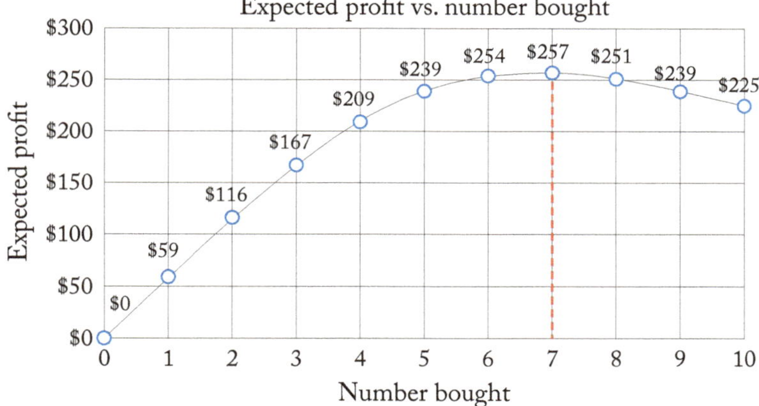

Figure 9.2: Expected profit as a function of the number bought in the sample problem.

$$\Pi(7) = -40 \cdot 7 + 100\{0 \cdot (0.01) + 1 \cdot (0.04) + \cdots + 7 \cdot (0.12) + 7 \cdot (0.12)\}$$
$$+ 25\{7 \cdot (0.01) + 6 \cdot (0.04) + \cdots + 1 \cdot (0.16)\}$$
$$= -280 + 483 + 54.25 = 257.25.$$

Figure 9.2 plots the expected profit as a function of the number of items bought. If we buy fewer than seven items, we lose money because we are likely to see more demand than the number we bought and so we are "leaving money on the table." If we buy more than seven, the likelihood of having to salvage items becomes very large and we are losing money by having too many on hand at the beginning of the sales season.

Figure 9.3 shows the marginal profit (revenue minus cost) as a function of demand when we buy seven items. Note that if the demand is 0 or 1, we lose money. If demand is 7 or more, we make $420. In these cases, we sell 7 items with a profit of $100 - 40 = 60$ on each item sold. If we multiply each number in this figure by the probability of that demand and sum over all demands, we get the expected profit of $257.25.

It turns out that we do not need to solve this problem using trial-and-error techniques. The optimal number to buy is given by the smallest value of B such that

$$Q(B) \geq \frac{p - c}{p - s}$$

where $\frac{p-c}{p-s}$ is called the *critical ratio*. In our case, $\frac{p-c}{p-s} = \frac{100-40}{100-25} = \frac{60}{75} = 0.8$. Using Table 9.2 we can see that the smallest value of demand such that the cumulative distribution is greater than 0.8 is 7, as found above. This is the optimal number to buy.

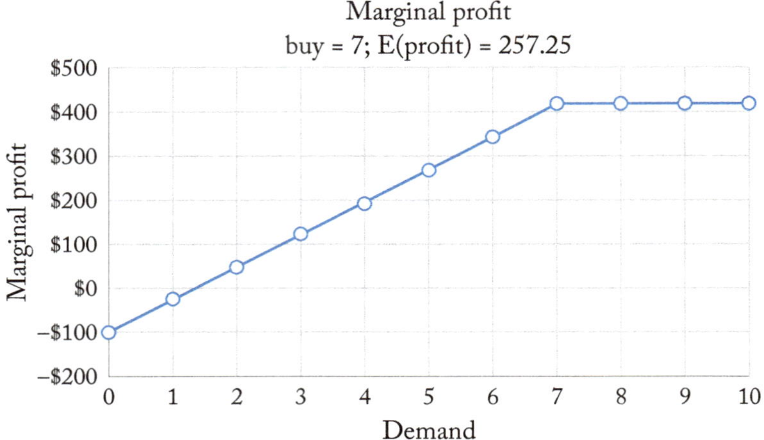

Figure 9.3: Expected revenue as a function of the demand when $B = 7$.

Table 9.3: Probability mass function and cumulative distribution of the number of burned-out bulbs each month

Number Out	Probability	Cumulative
0	0.02825	0.02825
1	0.12106	0.14931
2	0.23347	0.38278
3	0.26683	0.64961
4	0.20012	0.84973
5	0.10292	0.95265
6	0.03676	0.98941
7	0.00900	0.99841
8	0.00145	0.99986
9	0.00014	0.99999
10	0.00001	1.00000

9.3   A CONTRACTING EXAMPLE

We now turn our attention to an issue in contracting. Suppose the parking lot in your condominium complex is lit by 10 high-intensity lights. They tend to burn out. The probability that one will be burned out in any month is 0.3. The condominium association is planning to contract with either Liberty Lighting or Shining Signs to visit the parking lot once a month

Table 9.4: Terms of each contract

	Liberty	Shining
Overage cost	$175.00	$140.00
Base cost	$100.00	$115.00
Returned funds	$0.00	$20.00
Critical ratio	0.42857143	0.20833333
Optimal number	3	2
Expected cost	$398.06	$391.31
Cost at expected value	$300.00	$370.00

and to replace any burned-out bulbs. Table 9.3 gives the probability mass function and the cumulative distribution of the number of burned-out bulbs each month. The expected number of burned-out bulbs is three. Table 9.4 gives the terms of the contracts proposed by each company. Thus, Liberty Lights will charge $100 for each bulb for which your association contracts. If the number of burned-out bulbs is less than that number, no funds are returned. If more than that number are burned out, the firm will charge an additional $175 per bulb they replace. Similar values are shown for Shining Signs.

David, a member of the Association Board, argues that you should contract with Liberty since the average number of burned-out bulbs is three. With 3 burned-out bulbs, David argues, the cost would be $300 with Liberty and $345 with Shining. Edith, supports David's conclusion, but she argues that using the critical ratio method, you should contract for three bulbs with Liberty and only two with Shining. She computes the cost for Liberty to be $300 (an average of 3 burned-out bulbs times $100 per bulb under the contract). But, she says the cost with Shining will be $370. She gets this by multiplying the contracted 2 bulbs by $115 per bulb to get $230 for the contract cost, and adding to that the cost of the third bulb, $140, making the total equal to $370.

Tamar, however, believes that you should contract with Shining. She argues that you can't just look at the average number of bulbs that are burned out. You need to look *at everything that might happen*. She shows the Association Board Figure 9.4. She argues that when you multiply each cost by the probability that the given number of bulbs will be burned out, the expected cost for Liberty is $398.06. The expected cost using Shining would be only $391.31, an expected saving of $6.75.

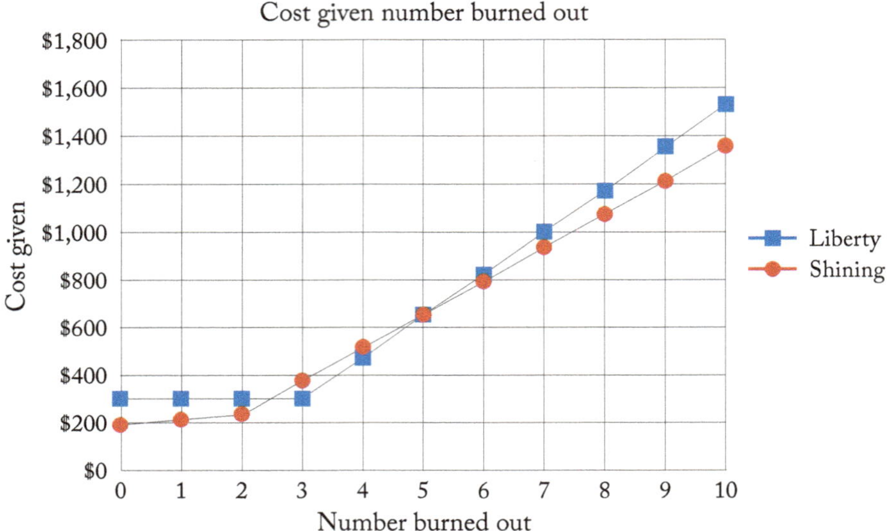

Figure 9.4: Cost under each contract for each number of burned-out bulbs.

In fact, Tamar is correct. You should contract with Shining and her reasoning is correct. The formula that you should use is given by

$$TC(B) = \underbrace{cB}_{\substack{\text{upfront} \\ \text{cost of} \\ \text{contract}}} + \sum_{d=0}^{10} q(d) \left\{ \underbrace{-s \cdot \max(0, B - d)}_{\substack{\text{refund if } d \text{ bulbs} \\ \text{are burned out}}} + \underbrace{p \cdot \max(0, d - B)}_{\substack{\text{extra cost if } d \text{ bulbs} \\ \text{are burned out}}} \right\}.$$

Again, the optimal number can be found using the critical ratio for each company. The critical ratios are shown in Table 9.4. It is worth noting that (1) this is another example of when the expected value of a function (as given in Figure 9.4) is not the same as the function evaluated at the expected value (or 3 in this case) and (2) the true expected costs for each company are significantly higher than either David or Edith suggested due to the costs associated with more bulbs than the average being burned out.

Finally, Table 9.5 gives the expected cost of contracting for each number of burned-out bulbs for each company. The optimal values for each company are shown in **bold red** letters. These values are also plotted in Figure 9.5. While the curves are relatively flat near the optimal values for each company, contracting for too many bulbs could be costly.

Table 9.5: Expected cost of each company for each number of bulbs under contract

Contract For	Liberty	Shining
0	$525.00	$420.00
1	$454.94	$398.39
2	$406.07	$391.31
3	$398.06	$412.24
4	$436.74	$465.19
5	$510.44	$542.16
6	$602.16	$631.48
7	$700.30	$725.21
8	$800.03	$820.02
9	$900.00	$915.00
10	$1,000.00	$1,010.00

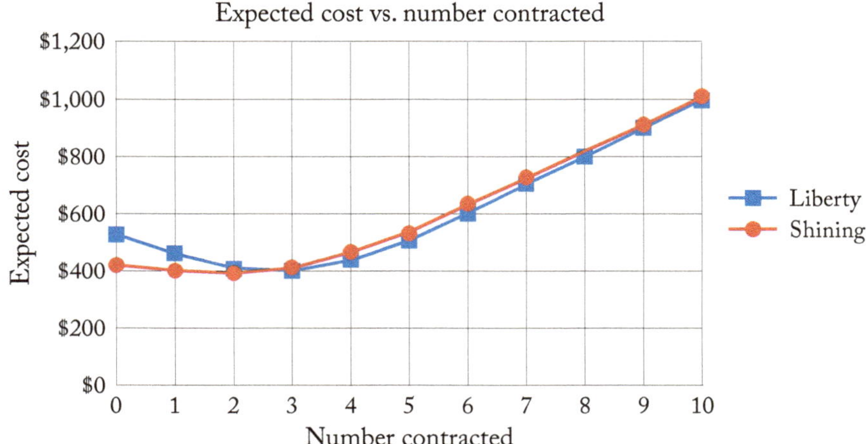

Figure 9.5: Expected cost of each company for each number of bulbs under contract.

C H A P T E R   10

Introduction to Linear Programming

10.1   WHAT IS LINEAR PROGRAMMING?

In Chapter 7, we introduced the four key elements of any optimization problem: inputs, decision variables, objective(s), and constraints. We showed how linear regression could be cast as an optimization problem. In Chapter 8, we used optimization to look at the tradeoff between fixed costs and holding costs in the economic order quantity model. In both those cases, the optimization problem could be solved using techniques from calculus. In particular, we simply take one or more derivatives and set the resulting equation(s) to 0 to solve for the optimal value(s) of the decision variables(s). In Chapter 9, we introduced demand uncertainty into the inventory problem. The derivation of the critical ratio entails taking a derivative (in the case of continuous demand functions) or looking at first-order differences (in the case of discrete distributions).

In this chapter we introduce one of the most powerful weapons in the arsenal of operations management: linear programming. Linear programming is at the heart of many optimization problems. In subsequent chapters we will show how to use linear programming and its extensions in a variety of problem contexts. At the end of this chapter, we will show how the newsvendor problem of Chapter 9 can be recast as a linear programming problem.

So, what is linear programming? An optimization problem is a linear programming problem if the objective function and the constraints can be written as **linear** functions of the decision variables. This means that we are not multiplying decision variables together or dividing one decision variable by another. We are not using trig functions, or raising a decision variable to some power, or taking the square root of a decision variable, or using exponentiation or logarithms in the objective function or in the constraints. Importantly, when using Excel, it means that decision variables do not appear in IF statements; IF statements are inherently nonlinear.

For our purposes now, any linear programming problem can be written using the notation outlined below. We are given a set of outputs that we want to produce (J) and a set of inputs (I) needed to produce those outputs. We know the value of each unit of output j (v_j), how much of each input i is available (b_i), and how much of each input i is needed to produce one unit of output j (a_{ij}). The problem is to determine how many units of each output j to produce (X_j) to maximize the total value of all items produced, subject to resource availability constraints. The meaning of outputs and resources will change depending on the problem context.

INPUTS

J set of outputs
I set of inputs
v_j value of one unit of output $j \in J$
b_i number of units of input $i \in I$ that are available
a_{ij} number of units of input $i \in I$ needed to make one unit of output $j \in J$

DECISION VARIABLES

X_j number of units of output $j \in J$ to produce

MODEL

$$\begin{aligned}
Max \quad & \sum_{j \in J} v_j X_j \\
s.t. \quad & \sum_{j \in J} a_{ij} X_j \le b_i \quad \forall i \in I \\
& X_j \ge 0 \quad\quad\quad\quad \forall j \in J .
\end{aligned}$$

The formulation of a generic linear programming problem is given above. $\sum_{j \in J} v_j X_j$ represents the total value of all outputs produced. The notation *s.t.* means *subject to*. This is followed by a set of inequalities (or equalities) that represent the constraints. For example, the left-hand side of the constraint $\sum_{j \in J} a_{ij} X_j \le b_i$ represents the total amount of resource $i \in I$ that is used in the production of all of the outputs. This quantity clearly has to be less than or equal to the total amount of resource $i \in I$ that is available, or b_i. This has to be true for every resource and the notation $\forall i \in I$ simply means for all values of the index i in the set of resources I. Finally, we require all output quantities to be non-negative with the constraints that state $X_j \ge 0$ for all outputs j in the set J.

It is important to note that any linear programming problem—whether the objective function is a minimization or a maximization; whether the constraints are less than or equal to constraints, greater than or equal to constraints, or equalities; and whether the decision variables must be non-negative, must be non-positive, or are unconstrained in sign—can be transformed to this model form. The mechanics of doing so are straightforward, but are beyond the scope of this text.

10.2 A GRAPHICAL VIEW OF LINEAR PROGRAMMING

In this section, we present a simple linear programming problem. Suppose you own a small bakery. You make pies and bread. You can sell as many pies and breads as you can produce. Your net profit on each pie is \$2.80 and your net profit on each loaf of bread is \$3.30. Your production is limited by three resources:

Table 10.1: Bakery example resources needed and available

	Pie	Bread	Available
Oven	0.35	1.05	7
Refrigerator	3	0.75	19
Labor	1	1.5	11.75

1. the amount of time you can operate your oven each day,

2. the number of cubic feet of refrigerator space you have, and

3. the number of hours the baker can work each day.

Table 10.1 shows the three resources and the required use of each of those in making a pie or a loaf of bread. For example, it takes 1 hour of labor to make a pie and 1.5 hours to make a loaf of bread, as shown in the last row. You have a maximum of 11.75 labor hours available each day. (Clearly, these are hypothetical numbers and do not necessarily reflect any real conditions. I have chosen them so that they illustrate key principles and concepts.)

With these inputs, we can now write down the linear programming problem for this hypothetical bakery as follows, where X is the number of pies to produce each day and Y is the number of loaves of bread to make:

$$
\begin{array}{llllllll}
Max & 2.8X & + & 3.3Y & & & & \text{profit} \\
s.t. & 0.35X & + & 1.05Y & \leq & 7 & & \text{oven time} \\
 & 3X & + & 0.75Y & \leq & 19 & & \text{refrigerator space} \\
 & X & + & 1.5Y & \leq & 11.75 & & \text{labor} \\
 & X & & & \geq & 0 & & \text{non-negativity} \\
 & & & Y & \geq & 0 & & \text{non-negativity}
\end{array}
$$

Note that we are not multiplying and decision variables by anything other than a constant.

While most companies will have multiple products—including chocolate chip cookies, chocolate muffins, and chocolate chip scones in the case of a local bakery—the advantage of focusing on only two outputs is that we can graph the problem. Figure 10.1 shows this problem graphically. The X-axis represents the number of pies we produce and the Y-axis represents the number of loaves of bread we make. These are the *decision variables*. We will begin by focusing on the constraints shown with solid green (oven time), orange (refrigerator space), and black (labor hours) lines. These are the three *constraints* in the problem. For a solution to be *feasible*, it must satisfy all of these constraints along with the *non-negativity* constraints. This means that it must lie within the five-sided region defined by the coordinates (0,0), (0,6.67), (3.5, 5.5), (5.25,4.33), and (6.33,0.0). This is called the *feasible region* and is shown in yellow in Figure 10.1.

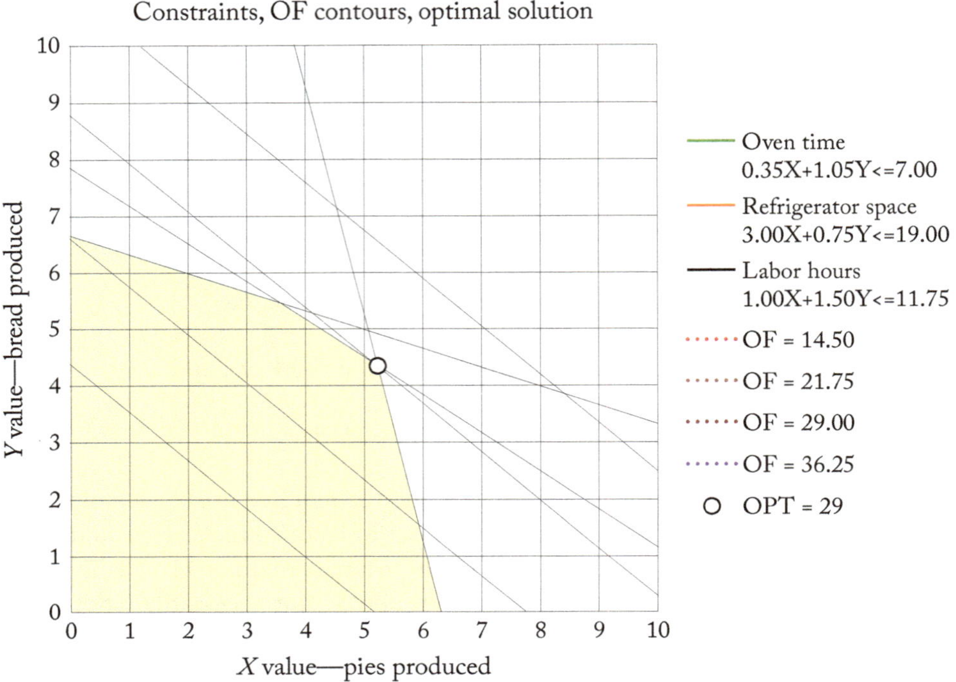

Figure 10.1: Constraints, feasible region, objective function (OF) contours, and optimal solution to the bakery problem.

The dashed lines in Figure 10.1 represent iso-contours of the objective function. On any such line, the objective function value is the same. Starting at the bottom, the orange dashed line shows solutions for which the objective function would equal \$14.50. We can clearly do better than this by pushing the line up and to the right. The next line represents those solutions with values equal to \$21.75. Again, we can do better. The third line represents solutions with a value equal to \$29.00. This line touches the feasible region at only one point. We cannot improve the objective function value any more. The fourth line represents solutions with an objective function value of \$36.25. While we would love to be operating on this line, or even a line with a larger value, there are no feasible solutions that correspond to this line. A grey circle shows the *optimal* solution. It occurs when we produce 5.25 pies and 4.33 loaves of bread each day. The objective function value at this point is 29. At least one optimal solution to any linear programming problem will occur at a "corner" point of the feasible region, or a point at which two or more constraints (possibly non-negativity constraints) intersect.

In the solution to this problem, the refrigerator space and labor hour constraints are the two constraints that limit production. They are called *binding* constraints. If we had just a little bit more of either one, we could generate more profit. For example, if we could work 1/4 of an hour more, changing the right-hand side of the labor constraint from 11.75–12.0, the optimal solution would have us making 5.2 pies and 68/15 or 4.533 loaves of bread. In this case, we decrease the number of pies slightly and increase the number of loaves of bread. The profit increases to $29.52. Note that if we had a little more or a little less oven time, the optimal solution would not change, since oven time is not a binding constraint. *Duality theory* in linear programming allows us to automatically compute the rate of change of the objective function for very small changes in the resources available to us. This, however, is beyond the scope of this introductory text.

Returning to the original problem with a constraint of 11.75 labor hours, we realize that making 5.25 pies and 4.33 loaves of bread each day is not realistic. We need to be making an integer number of pies and loaves of bread. We can add a constraint to the problem that says that X and Y must be integer-valued. This changes the problem from a pure *linear programming* problem to an *integer linear programming* problem. In general, integer problems are harder to solve, though for small problems like this we can solve the problem very easily.

Figure 10.2 shows this problem. The feasible region now consists of the solutions shown as blue dots. The optimal solution is now to produce 4 pies (or 2^2 pies) and 5 (a prime number) loaves of bread. The optimal profit is now $27.70. From this we can learn that, while bread is a prime ingredient of any sandwich, pies are squared.

Seriously, we can learn two important lessons. First, rounding the linear programming solution to the nearest integer solution may not be optimal. In fact, it may not even be feasible. In our case, we would have rounded the solution to 5 pies and 4 loaves of bread. The profit in that case would have been $27.20, or $0.50 less than the optimal profit of $27.70 found when we produce 4 pies and 5 loaves of bread. Second, we note that the profit has decreased from the profit found in the pure linear programming solution. Whenever we add a constraint to a maximization problem, the objective function value will either stay the same (if we are lucky) or go down (which is more common). Similarly, if we have a minimization problem, adding a constraint will either keep the objective function value the same (if we are lucky) or force it to increase (which is more likely).

Integer programming can be—and often is—a graduate course in its own right and is beyond the scope of this text.

10.3   THE NEWSVENDOR PROBLEM AS A LINEAR PROGRAMMING PROBLEM

Finally, we show that the newsvendor problem of Chapter 9 can also be formulated as a linear programming problem. We modify the notation of Chapter 9 slightly as shown below. In particular, we let q_d be the probability that demand is exactly d units and we assume that the maximum demand with a non-zero probability is $d = n$.

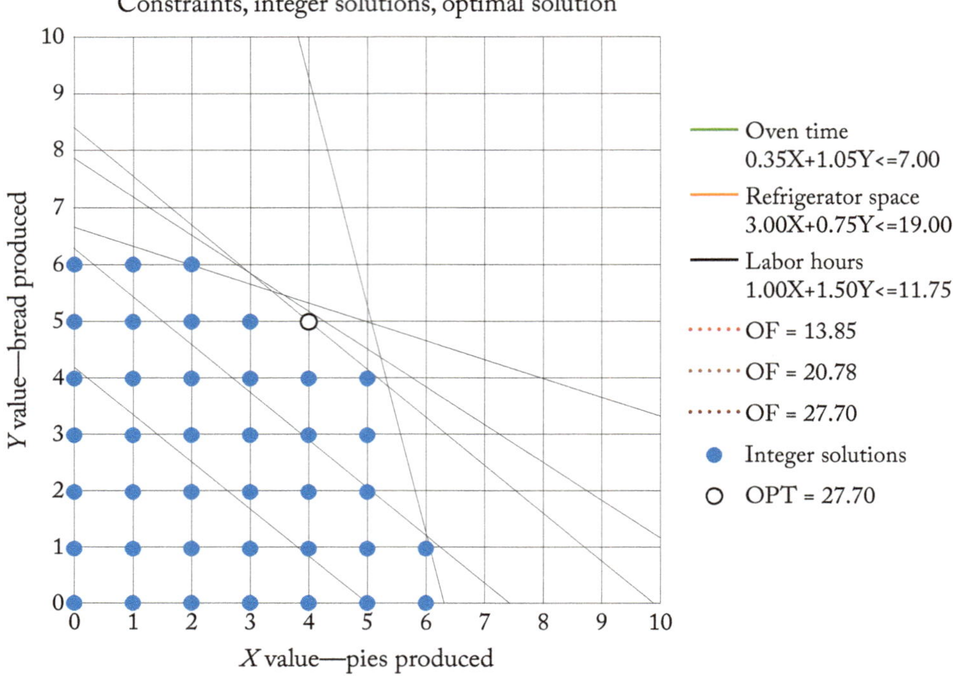

Figure 10.2: Bakery problem requiring an integer-valued solution.

INPUTS

p unit sale price

c unit cost

s unit salvage value

D a random variable representing the demand

q_d probability that the demand is exactly d units

n maximum demand with a non-zero probability

We define the following decision variables:

DECISION VARIABLES

B number to buy

X_d number sold if demand is equal to exactly d units

Y_d number salvaged if demand is equal to exactly d units

Note that for each possible demand value (or demand scenario), we have a separate variable denoting the number of items sold in that demand scenario, X_d, and the number of items salvaged in that demand scenario, Y_d. With this notation, the problem can be formulated as follows:

MODEL

$$
Max \quad \underbrace{-cB}_{\substack{\text{cost of} \\ \text{items bought}}} \quad + \underbrace{\sum_{d=0}^{n} q_d \left(\underbrace{pX_d + sY_d}_{\text{revenue if demand} = d} \right)}_{\text{expected revenue over all demand values}}
$$

$$
\begin{array}{llll}
s.t. \quad & X_d & \leq d & d = 0, \ldots, n \quad \text{sales} \leq \text{demand} \\
& X_d & \leq B & d = 0, \ldots, n \quad \text{sales} \leq \text{number available} \\
& X_d + Y_d & = B & d = 0, \ldots, n \quad \text{sell or salvage everything} \\
& X_d & \geq 0 & d = 0, \ldots, n \quad \text{non-negativity} \\
& Y_d & \geq 0 & d = 0, \ldots, n \quad \text{non-negativity} \\
& B & \geq 0 & \quad \text{non-negativity}
\end{array}
$$

The first term of the objective represents the cost of the items we choose to buy. The second term captures the expected revenue, which is composed of the revenue from sales plus the revenue from salvaged items in each demand scenario multiplied by the probability that the demand equals d. The first constraint states that no matter what we buy, we cannot sell more than the demand. The second constraint says that no matter what the demand, we cannot sell more than the number of items we buy. The third major constraint says that every item we buy is either sold or salvaged. Finally, we have the non-negativity constraints.

This linear programming problem is also an example of a *stochastic programming* problem. We must make some decisions now, before uncertainty is revealed. In this case, we must decide how many items to buy. This is referred to as the *first-stage* decision. After uncertainty is revealed (after we know what the demand is), we can decide on *scenario-specific* decisions including, in this case, how many items to sell and how many to salvage. Stochastic programming is a very active area of research and application today as managers and decision makers attempt to deal with uncertainty [Louveaux and Birge, 2011].

Finally, we note that this problem can also be represented as a *network* flow problem. A network flow problem is a special form of a linear programming problem that can be represented using nodes and arcs. At every node, we require the flow in to equal the flow out. Associated with each arc is a lower bound, or the minimum amount of flow required on the arc, a unit cost of the flow on the arc, and an upper bound, or the maximum amount of flow allowed on the arc. The problem is to find the set of flows that minimize the total cost while obeying the lower

Table 10.2: Example probability mass function for demand

Demand	Probability
0	0.10
1	0.15
2	0.25
3	0.25
4	0.15
5	0.10

and upper bounds on every arc and ensuring that the flow into every node equals the flow out of every node.

To illustrate the way in which the newsvendor problem can be formulated as a network flow problem, consider a problem with the probability mass function for demand as given in Table 10.2. q_d is the probability that the demand is exactly d units and Q_d is the probability that the demand is d or fewer units. Associated with each arc are three numbers: the lower bound on the flow on the arc, the unit cost of flow on the arc, and the upper bound on the flow on the arc.

Figure 10.3 shows the resulting network flow problem. Figure 10.4 shows the resulting network when the sale price of the items is $p = 225$, the unit cost is $c = 125$ and the salvage value is $s = 20$. The total number of nodes will be $n + 1$, where n is the largest possible demand.

The top red arcs represent buying items from the supplier at a cost of c. The bottom green arcs represent the expected revenue from sales and from salvaging items that results from the nth item that was purchased. Thus, for example, for the first item that is purchased, we can sell it with probability $1 - Q_0$ resulting in an expected sales revenue of $p(1 - Q_0)$ and we will have to salvage it with probability Q_0, resulting in an expected salvage revenue of $s Q_0$. The total expected revenue from the first item we buy is therefore $p(1 - Q_0) + s Q_0$. The negative of this revenue is the "cost" associated with the arc from node 1 to node 0. Other green arcs have similar unit costs.

It should be clear from either Figure 10.3 or Figure 10.4 that we will want to incur the cost of buying an additional item, c, if that cost is more than offset by the expected revenue we obtain from buying that item. In other words, we will want to buy the item B if $c < p(1 - Q_{B-1}) + s Q_{B-1}$. This means that we buy the Bth item if $Q_{B-1} < \frac{p-c}{p-s}$, or we find the smallest value of B, such that $Q_B \geq \frac{p-c}{p-s}$. But this is just the critical ratio optimality condition that we outlined in Chapter 9.

Figure 10.5 shows the optimal solution to the example newsvendor problem where the first number on each arc is the optimal flow and the second is the unit cost in this case. The objective function value is -128.25, which is the negative of the optimal expected profit.

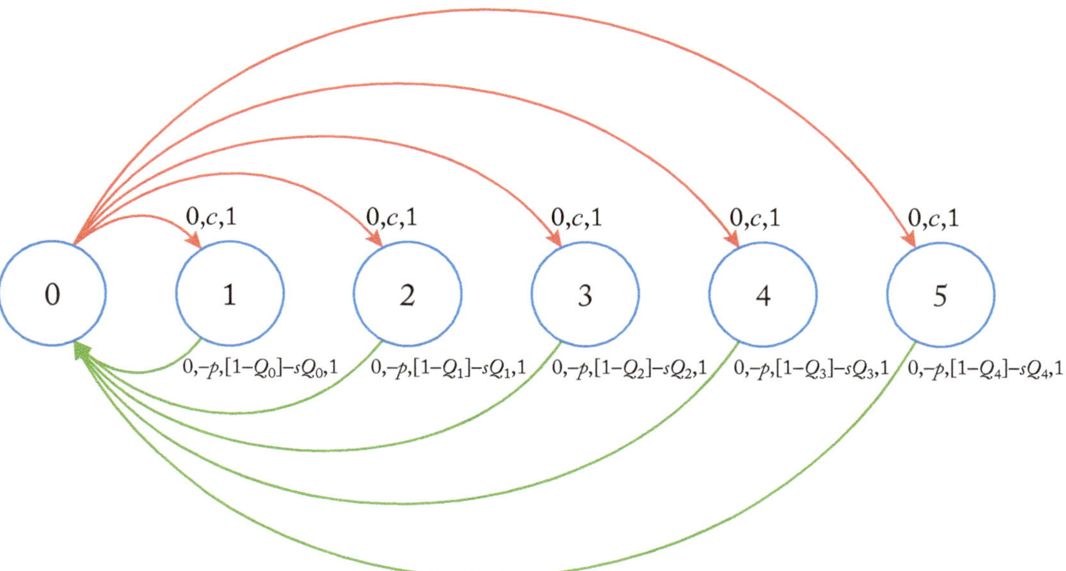

Figure 10.3: Network flow representation of a newsvendor problem.

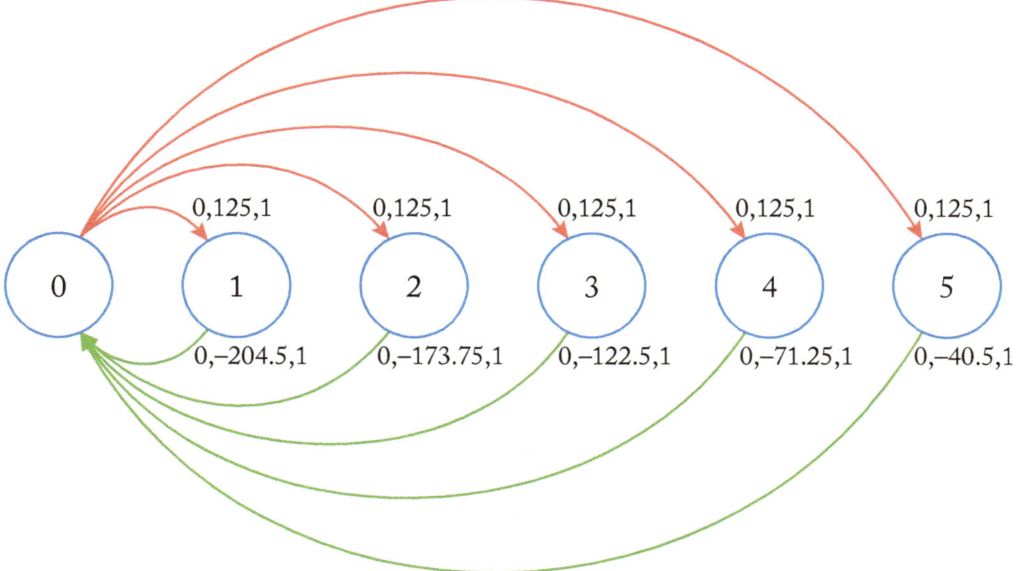

Figure 10.4: Network flow when price = 225, cost = 125, and salvage = 20.

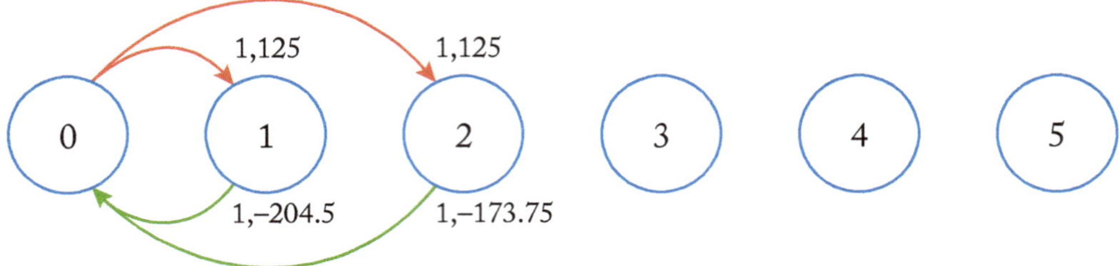

Figure 10.5: Optimal solution to the example newsvendor problem.

So why do we care about being able to represent and solve the problem as a network flow problem? There are a number of reasons. First, such representations often make it easier to gain insights into the problem. Second, specialized algorithms solve network flow problems faster than more general linear programming algorithms can solve such problems. In fact, many important real-life problems are network flow problems. For example, when you use a mapping program to route you from your origin to your destination, the underlying algorithm is a shortest path algorithm, which is a form of network flow problem. Third, it may be possible to build this network structure into more advanced models in which the network flow problem is only one component of the overall model.

The next two chapters illustrate the use of linear programming in two important contexts. Chapter 11 deals with optimal repositioning of vehicles (e.g., bicycles or scooters) for ride-sharing systems. Chapter 12 deals with assigning students to seminars to optimize satisfaction.

10.4 REFERENCES

F. Louveaux and J. R. Birge, *Introduction to Stochastic Programming*, 2nd ed., Springer, New York, 2011. DOI: 10.1007/978-1-4614-0237-4. 69

CHAPTER 11

The Transportation Problem

11.1 ONE OF MANY PROBLEMS

In Chapter 10, we introduced linear programming. Linear programming, and its many extensions, is an exceptionally powerful tool that can and has been used to address numerous problems in operations management. A very partial list of those problems includes:

1. Finding the shortest path from one point to another. This is done millions of times every day by users of GPS and mapping software.

2. Determining when to replace a machine. The machine might be your own car or the office copier or a multi-million dollar aircraft.

3. Assigning medical residents to rotations throughout the year.

4. Scheduling workers in an emergency room that faces time-varying demands or loads.

5. Assigning students to sections to maximize student diversity within each section.

6. Relocating empty rental vehicles at the end of a day.

7. Assigning students to seminars to maximize overall student satisfaction.

8. Locating emergency medical service vehicles to ensure adequate coverage of a city.

9. Locating fulfillment centers for Amazon or any other large online retailer.

10. Selecting suppliers for an auto manufacturer in the face of demand uncertainty and exchange rate uncertainty.

11. Locating emergency supplies in advance of potential natural disasters including earthquakes.

The list could easily go on and on. In this chapter and the next we present two examples of linear programming related to examples (6) and (7) above. We begin with the problem of relocating empty rental vehicles at the end of a day.

11.2   THE TRANSPORTATION PROBLEM

Let us consider the problem faced by a bicycle rental company. Early each morning, people leave their homes for their offices or other places of employment. Some will use a local bike rental agency. They may walk a short distance to a bike rack, use their mobile device to unlock a bike, and then ride to their destination. At the end of the day, of course, some may reverse the trip. Some, however, may have evening plans close to their workplace. They may go to a business dinner with a new employee; they may have a personal date; they may go to dinner and a movie with a colleague or friend; or they may go shopping. In any event, they may *not* rent a bike for their after-work activities, preferring instead to use a taxi or to travel with a friend or colleague in a private vehicle.

Those who use the bicycle for a one-way trip create a potential problem for the rental company since there will then be an imbalance between where bicycles are left at the end of the day and where they need to be the next morning in anticipation of the rental cycle beginning anew. *Empty vehicles must be repositioned.* The rental company must move some bicycles from locations (or bike racks) with an oversupply of bicycles at the end of the day to locations with a deficit or demand for bicycles the following morning. This is an example of what is called the *transportation problem*.

Figure 11.1 illustrates this problem using the Census tract map of Ann Arbor, Michigan. The city has a population of about 115,000 residents according to the 2010 Census. It is divided into 33 Census tracts with an average population per tract of nearly 3,500 people. The tract populations range from a low of about 1,550 to a high of 7,100. The city is dominated by the University of Michigan's Ann Arbor campus, which is located roughly in the middle of the city. In Figure 11.1, green dots represent tracts that have an oversupply of rental vehicles at the end of the day. They have a *supply* of vehicles. Red dots represent tracts that have a *demand* for rental vehicles the next day. Note that the example data were generated so that the supply nodes would be near the downtown area, reflecting the sort of rental behavior outlined above.

Associated with every node is the actual supply or demand at that node or in the tract. For example, a green node might have a supply of 31 bicycles and a particular demand node might have a demand for 8 vehicles. In addition, we would know the distance between each pair of nodes. The problem is then to figure out how to move bicycles from their current (green) locations to where they need to be the following morning (the red nodes).

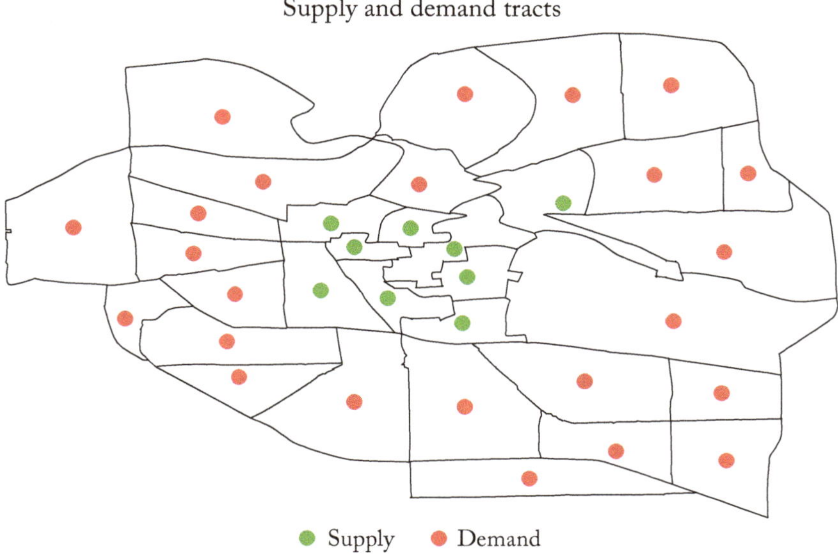

Figure 11.1: Example of supply and demand locations in Ann Arbor, MI.

Let us define the following notation:

INPUTS

J	set of supply nodes
K	set of demand nodes
s_j	supply of rental vehicles at node $j \in J$
h_k	demand for rental vehicles at demand node $k \in K$
d_{jk}	unit cost of moving a vehicle from supply node $j \in J$ to demand node $k \in K$

DECISION VARIABLES

Y_{jk}	number of vehicles to move from supply node $j \in J$ to demand node $k \in K$

With this notation, we can formulate a linear programming problem that minimizes the cost of relocating vehicles from supply nodes to demand nodes. We cannot move more vehicles than are available at any node.

$$Min \quad \sum_{j \in J} \sum_{k \in K} d_{jk} Y_{jk} \qquad\qquad \text{redistribution cost}$$

$$s.t. \quad \sum_{k \in K} Y_{jk} \leq s_j \qquad \forall j \in J \qquad \text{can not move more than is available}$$

at each supply node

$$\sum_{j \in J} Y_{jk} \geq h_k \qquad \forall k \in K \qquad \text{must satisfy demand at each demand node}$$

$$Y_{jk} \geq 0 \qquad \forall j \in J, \ k \in K \quad \text{movements must be non-negative}$$

The objective function minimizes the total redistribution cost. This is just the cost of moving vehicles from supply node $j \in J$ to demand node $k \in K$ given by the unit cost of such movements times the number of vehicles we move, $d_{jk} Y_{jk}$, summed over all demand and supply nodes. The first constraint says that we cannot ship more vehicles from supply node $j \in J$ than we have at that location. The summation represents the total number of vehicles, or bicycles in our case, that we move from supply node $j \in J$ to all demand nodes. The next constraint is the demand constraint and it says that we must satisfy the demand needs of every demand node. The summation on the left-hand side of the constraint represents the total number of vehicles moved to demand node $k \in K$ from all supply nodes. Finally, the last constraint says that we must have non-negative flows or vehicle movements.

Table 11.1 gives sample data for the 33 Census tracts of Ann Arbor. The first 9 rows correspond to supply nodes since they have a negative net demand and the last 24 rows correspond to demand nodes with a positive net demand. The distances or costs of moving vehicles between nodes are given by the great circle distances between the node centroids, or simply by the straight line distance between the nodes.

Figure 11.2 shows the optimal solution to this problem. There are a number of things to notice. First, flows go from green supply nodes to red demand nodes as expected. Second, there are some nodes or tracts that ship to only one other tract. In fact, there are two such tracts, both of which are in the southeast part of the downtown area. One tract near the south central part of the city ships to 6 other tracts. Third, most demand tracts can have their requirements met by shipments from only one supply node. This is not surprising since there are roughly 2.5 times as many demand nodes in this example as there are supply nodes. One tract in the southeast part of the city, however, requires shipments from three supply nodes. Fourth, there are exactly 32 non-zero flows in this solution. In fact, we can show that if there are n nodes (the total number of supply and demand nodes), there will be at most $n-1$ non-zero flows. Also, there are no cycles, meaning that if we look at the network formed by the set of flows, there is at most (in this case exactly) one way to go from any node to any other node, ignoring the directionality of the arrows in the figure. This is true of any solution to any transportation problem. Finally, we note that the total cost of this solution is 267.871, meaning that we would have that many bicycle-miles of movements. This is the distance associated with each blue arrow in Figure 11.2,

Table 11.1: Sample supply and demand data for the Ann Arbor Problem

Census Tract	Flow In	Flow Out	Net Demand	Census Tract	Flow In	Flow Out	Net Demand
4005	155	119	−36	4046	111	116	5
4002	185	154	−31	4033	107	113	6
4022	134	115	−19	4036	131	137	6
4007	78	61	−17	4043	80	86	6
4006	117	101	−16	4054	66	73	7
4008	71	59	−12	4021	84	92	8
4001	48	37	−11	4034	80	88	8
4003	167	161	−6	4035	76	84	8
4004	80	77	−3	4055	75	83	8
4026	60	61	1	4056	113	121	8
4041	91	92	1	4027	148	157	9
4042	57	58	1	4032	92	101	9
4051	57	58	1	4038	86	95	9
4031	50	53	3	4045	112	121	9
4023	71	75	4	4052	125	138	13
4025	55	59	4	4053	129	142	13
4044	57	61	4				

multiplied by the number of bicycles moved from the corresponding supply node to the demand node, summed over all such arrows.

11.3 THE TRANSPORTATION PROBLEM AS A NETWORK FLOW PROBLEM

The transportation problem is actually a network flow problem. Recall that a network flow problem can be represented as a problem with nodes and arcs. Associated with each arc is the lower bound or minimum required amount of flow on the arc, the unit cost of using the arc, and an upper bound or the maximum allowed flow on each arc. We require the flow into each node to equal the flow out of each node and for the flows to be between the lower and upper bounds on each arc. The problem is to find the minimum cost set of flows that satisfy these conditions.

Figure 11.3 shows the example Ann Arbor Transportation Problem as a network flow problem. The Start and End nodes are shown in blue, supply nodes are shown in green, and

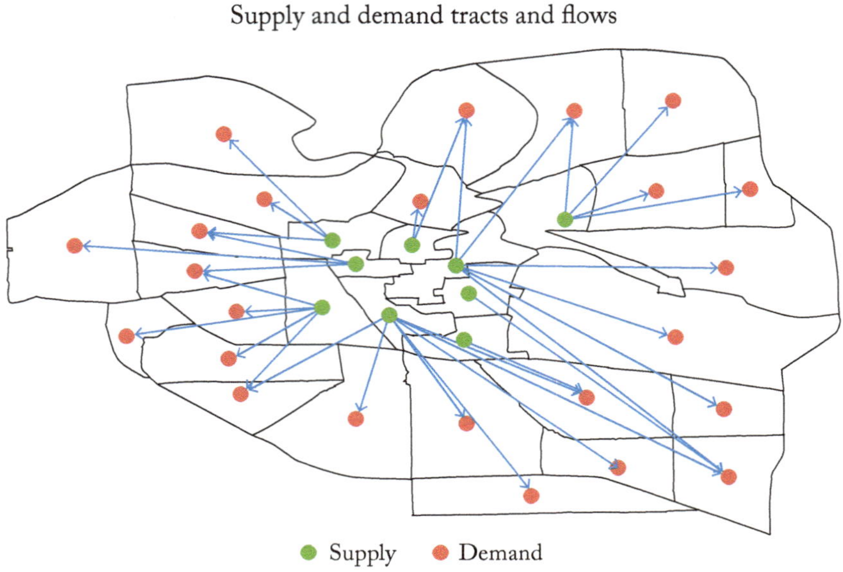

Figure 11.2: Optimal flows for the Ann Arbor Problem.

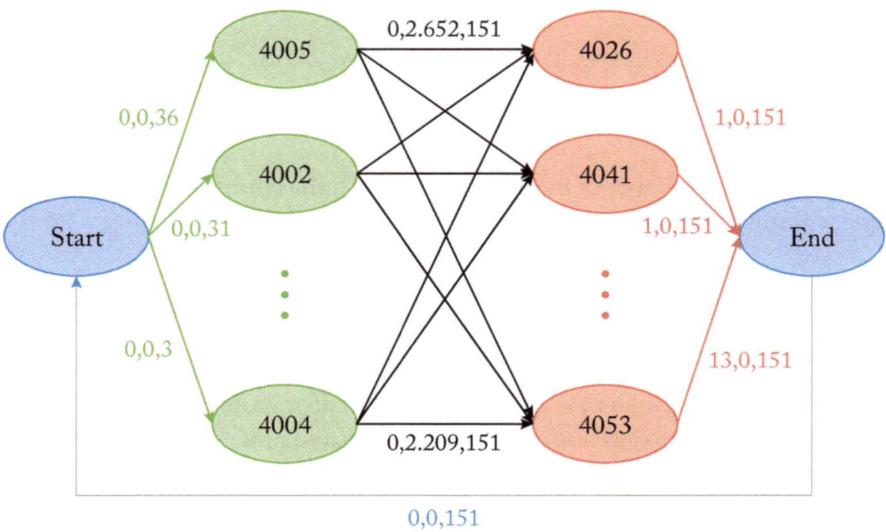

Figure 11.3: The Ann Arbor Transportation Problem.

demand nodes are shown in red. Nodes are listed in order of descending supply and ascending demand, corresponding to the order shown in Table 11.1.

The lower bound and unit cost on each arc from the start node to each demand node is 0, while the upper bound on each such (green) arc is the supply at the corresponding node. The lower bound on each (black) transportation arc from a (green) supply node to a (red) demand node is 0. The unit cost on each such arc is the distance between the two nodes, measured in miles in this case. The upper bound is 151, which is the total number of vehicles that must be relocated. For the (red) arcs between the demand nodes and the End node, the lower bound is the required flow into the demand node. The unit cost is 0 and the upper bound is again 151. For the return arc from the End node to the Start node, the lower bound is 0, the unit cost is 0 and the upper bound is 151. The lower bounds on the (red) demand arcs force there to be flow in the network.

Again, it is useful to recognize that a problem can be structured as a network flow problem as there are very efficient and fast algorithms for solving such problems. Also, there are some theoretical properties of the solution to such problems that are also useful, though they are beyond the scope of this text.

11.4 WHAT IS WRONG WITH THIS MODEL?

As noted in Chapter 1, George Box once said, "All models are wrong, but some are useful." It is always a good idea to ask what is wrong with this model. We illustrate that with the use of the transportation problem formulation for the problem of relocating rental vehicles at the end of the day.

There are two key questions to ask in doing this sort of analysis or critique: (1) What assumptions does the model make and are they right? (2) What is left out of the model that may be important or relevant to the problem at hand?

One of the key assumptions that we made is that there is a *unit* cost associated with relocating each bicycle from a supply node to a demand node. In fact, bicycles are likely to be loaded onto a truck at a supply node and then trucked to the relevant demand nodes. The cost of doing so is not likely to depend linearly on the number of bicycles moved from one location to another since moving one bicycle costs almost as much as moving 10 bicycles. Second, and related to this assumption, is the assumption that bicycle moves occur directly from a supply node to a demand node. In fact, the truck may load 25 bicycles at a time and then make several stops at demand nodes. In other words, bicycles are likely to be delivered to demand locations using a *route* and not an *out-and-back* trip. Thus, the model suffers from a key assumption that may make the results less than representative of any real world relocation problem for small rental vehicles like bicycles or scooters.

One facet of the real world that is missing from this problem is *uncertainty*. The model assumes that we know how many bicycles will be needed at each location the following morning. Unless all rentals are to customers with a long-standing contract with the rental company, the demand at any location or in any tract the following morning is likely to be uncertain. Therefore, an important question is how many bicycles should the company deliver to any location given that the demand is uncertain and there are costs associated with not having enough bicycles on hand for rental during the day and there is a cost of having too many bicycles at any location. This aspect of the problem should be reminiscent of the newsvendor problem. Finally, we need to be concerned not only with balancing the demand and supply of bicycles at the beginning of the day, but throughout the day. In fact, operations management techniques can be used to address all of these issues. This is an active area of research, as there has been a proliferation of mobility rental companies around the world, ranging from auto rental firms to bicycle rentals to scooter rental companies.

Other key questions that this model ignores are: (1) how many bicycle rack locations should there be in each tract and where should they be and (2) how many bicycles should each rack accommodate. Questions like these can be addressed, in part, using location models as discussed in Chapters 14 and 15 below.

In short, this model may provide some insights into this problem. A major auto manufacturer has used this model, with appropriate extensions, in determining how to relocate empty wire bins used to transport stamped metal parts from fabrication plants to assembly plants. That said, it is clear that there are important questions associated with the management of rental vehicles that this model does not adequately address.

C H A P T E R   12

Assigning Students to Seminars

12.1   ORIGIN OF THE PROBLEM

Many years ago, while I was still a faculty member at Northwestern University, I received a call from an associate dean of the Weinberg College of Arts and Sciences at the university. She told me that in the Fall term each of the 1,000 entering first-year students takes a seminar. There were about 70 seminars and each student rank ordered his/her top 20 seminars. She then spent two weeks working half time each day assigning students to seminars. Some students ended up being assigned their 15th choice seminar, or worse. Imagine that letter in today's terms: "Welcome to Northwestern University. You were assigned to your 15th choice seminar. By the way, you owe us $56,691 in tuition and fees for the year." This is not very welcoming. She wondered if there was a better way of doing this assignment. Within a week, I had software running on her desktop computer that took seconds to give her three or four alternate assignments. Typically students were assigned to one of their top three choices. How did we do it? That is the story of this chapter.

12.2   ANOTHER LINEAR PROGRAMMING PROBLEM

It turns out that assigning students to seminars is very similar to the problem of relocating rental vehicles that we addressed in Chapter 11. In the course of this chapter, we will also show that some optimization results are not immediately intuitive.

To see how to formulate this problem, let us define the following notation:

INPUTS

J set of seminars

K set of students

s_j capacity of seminar $j \in J$

r_{jk} rank ordering of seminar $j \in J$
       by student $k \in K$ (1 is top choice)

DECISION VARIABLES

$$Y_{jk} \quad \begin{cases} 1 & \text{if student } k \in K \text{ is assigned to seminar } j \in J \\ 0 & \text{if not.} \end{cases}$$

With this notation, the model can be formulated as shown below:

$$\begin{array}{lll} Min & \displaystyle\sum_{j \in J} \sum_{k \in K} r_{jk} Y_{jk} & \text{total assigned rank} \\ s.t. & \displaystyle\sum_{k \in K} Y_{jk} \leq s_j \quad \forall j \in J & \text{must obey seminar capacities} \\ & \displaystyle\sum_{j \in J} Y_{jk} = 1 \quad \forall k \in K & \text{each student must be assigned to a seminar} \\ & Y_{jk} \in \{0, 1\} \quad \forall j \in J,\, k \in K & \text{binary assignments} \end{array}$$

The objective function minimizes the sum of all student assignments. If student 1 is assigned to her second choice seminar, she contributes two to the objective function value. If student 2 is assigned to his third choice seminar, he contributes three. If student 3 is assigned to her first choice seminar, she contributes one, and so on. If we divide the objective function by the number of students, we would get the average ranking for all assignments. The first constraint says that we cannot assign more students to a seminar than the capacity of the seminar. The second constraint says that each student must be assigned to exactly one seminar. Finally, the decision variables must be either 0 or 1. Actually, as we will show below, this problem can also be structured as a pure linear programming problem and as a network flow problem. Therefore, we could replace the final constraint with a simple non-negativity constraint as follows:

$$Y_{jk} \geq 0 \quad \forall j \in J,\, k \in K \quad \text{non-negativity.}$$

Table 12.1 shows sample rankings for a problem with 4 seminars and 16 students. The seminars are in (1) Computer Science, (2) Operations Research, (3) European History, and (4) Number Theory. The last row of the table shows the average ranking of each seminar. Not surprisingly, Operations Research is the most popular seminar; this is an operations management book after all. Number Theory is the least popular seminar.

If each seminar can accommodate exactly four students, we will obtain the optimal assignments shown in Table 12.2. All but two students, Gary and Lauren, are assigned to their first choice seminar and those two students are assigned to their second choice seminar. The objective function value is 18 resulting in an average assigned ranking of 1.125.

Table 12.3 summarizes this result and three other cases. The table contains four columns. The first describes the conditions of the model. To reflect some of these conditions, we would require an extension of the basic model outlined above. For example, we would need to determine which seminars to offer and which seminar to allow to exceed its nominal capacity. Additional constraints would be needed to reflect these conditions and to incorporate these additional de-

Table 12.1: Sample rankings for a small problem

Name	Comp Sci	Operations Research	European History	Number Theory
Alice	2	3	4	1
Babette	2	4	3	1
Carol	4	2	3	1
David	4	2	3	1
Eunice	1	4	3	2
Faith	2	3	1	4
Gary	1	3	2	4
Harry	4	2	1	3
Ingrid	3	1	2	4
Jonah	2	1	3	4
Keren	2	1	3	4
Lauren	3	1	2	4
Margery	2	1	3	4
Naomi	1	2	3	4
Oscar	1	2	3	4
Paul	1	2	3	4
Average	2.1875	2.125	2.625	3.0625

Table 12.2: Optimal assignment with seminar capacities of four

Student	Assign to	Assigned Ranking
Alice	Number Theory	1
Babette	Number Theory	1
Carol	Number Theory	1
David	Number Theory	1
Eunice	Comp Sci	1
Faith	European History	1
Gary	European History	2
Harry	European History	1
Ingrid	Operations Research	1
Jonah	Operations Research	1
Keren	Operations Research	1
Lauren	European History	2
Margery	Operations Research	1
Naomi	Comp Sci	1
Oscar	Comp Sci	1
Paul	Comp Sci	1
Total		18

Table 12.3: Summary of assignment results for other cases

Case	Offered	Extra	Objective Function
Base case; 4 sections of 4 students each	Comp Sci; Operations Research; European History; Number Theory		18
Allow one section to exceed capacity by 1	Comp Sci; Operations Research; European History; Number Theory	Operations Research	17
Offer 3 seminar of 5 students each with one seminar exceeding capacity by 1 student	Comp Sci; Operations Research; Number Theory	Operations Research	19
Offer 3 seminars of 5 students each, allow one to exceed capacity, but eliminate Number Theory which is the lowest ranked	Comp Sci; Operations Research; European History	Comp Sci	23

cision variables. This modeling is left as an exercise for the reader. The second column identifies the actual seminars to be offered. The third column shows which seminar is allowed to exceed its capacity by one student. The final column shows the value of the objective function.

In the second row, we consider the case in which each seminar has a nominal capacity of four students, but we allow one seminar to exceed its capacity by one student. It turns out, not surprisingly, that we expand the Operations Research seminar. European History only gets three students. By allowing for this expansion of a seminar, we are, in effect, relaxing a constraint. The objective function improves to 17 or an average assigned ranking of 1.0625. All but one student, Gary, is assigned to his or her first choice seminar.

In the third row, we ask what if we only offered three seminars, each with a capacity of five students, and allowed one seminar to exceed its capacity by one student. It turns out that we should eliminate European History and allow one extra student in the Operations Research seminar. The total assigned ranking goes up to 19 for an average ranking of 1.1875.

In some sense, the result shown in the third row might be surprising. We might have thought that we should eliminate Number Theory since it had the worst average ranking. In the fourth row, we show the results of doing just that and again having three seminars each with a capacity of five students and allowing one seminar to exceed its capacity. The total assigned ranking is now 23 for an average ranking of 1.4375, a significant degradation over the result shown in the third row. Computer Science is now the seminar with added capacity. It turns out that we could add one student to Operations Research, the most popular seminar in this case, and also end up with a total ranking of 23.

Intuitively, the reason it is better to leave Number Theory in the list of offered seminars even though it has a worse *average* ranking than European History is that the *average* is based on the ranking assigned by every student to each seminar. In none of the four cases of Table 12.3 is a student assigned to his/her fourth choice seminar. Including these choices in the average in Table 12.1 turns out to be somewhat misleading since no one is ever assigned to their fourth choice seminar. In fact, even in row four, only two students are assigned to their third choice seminar; all others are assigned to either their first or second choice seminars.

It is worth noting that the problem for Northwestern University had roughly 70,000 variables. Again, in about 5 seconds, we could identify several different solutions that differed in the average assigned value and the worst-case assigned value for the students.

12.3   STUDENT ASSIGNMENT AS A NETWORK FLOW PROBLEM

This problem can also be represented as a network flow problem. Figure 12.1 shows how we can do this for the problem in the first row of Table 12.3. The network looks very similar to the transportation problem network of Chapter 11. Again, we have a Start and End node. There is one (green) node representing each seminar. The capacity of each seminar is captured in the upper bound on the flow from the Start node to each seminar node. Each student is represented

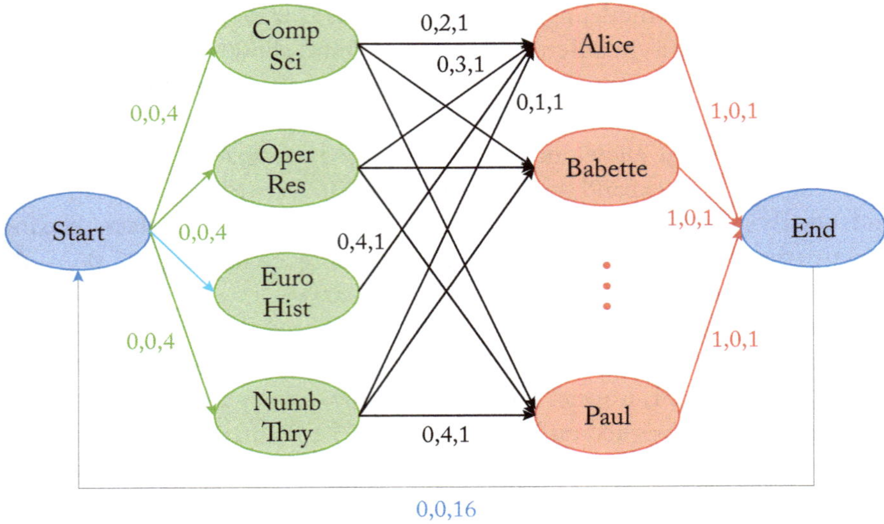

Figure 12.1: Student assignment as a network flow problem.

by a (red) node, which is connected to the End node. The lower and upper bounds on these arcs are both 1 meaning that each student must be assigned exactly once. The (black) arcs connecting seminar nodes to student nodes each have a lower bound of 0, an upper bound of 1, and a unit cost equal to the ranking that the student gave to that seminar. Only a subset of these arcs and arc values are shown. Alice, for example, ranked Computer Science as her second choice, Operations Research as her third choice, European History as her fourth choice, and Number Theory as her first choice. Paul ranked Number Theory as his fourth choice seminar.

One of the properties of a network flow problem of this sort is that the flows will naturally be integer-valued as long as the lower and upper bounds are integers. Since the latter condition is met in this case, we can be assured that solving this problem as a network flow (or linear programming) problem will result in a solution that automatically satisfies the binary condition on the assignment variables, Y_{jk}. It was through representing the problem as a series of larger and larger network flow problems, progressively including arcs corresponding to higher student rankings of seminars, that we were able to (1) solve the large Northwestern problem quickly and (2) provide the associate dean with a number of different solutions to the problem.

CHAPTER 13

Analytic Location Modeling

13.1 OVERVIEW OF LOCATION MODELING

Location modeling is one of the important topics in operations management. If a firm or service operation gets its locations wrong, almost all other costs are likely to be higher than necessary. Facility location modeling deals with such questions as follows.

1. Where should an online retailer, like Amazon, have fulfillment centers, how many should there be, how large should each be, and what customers should be in the primary catchment area of each fulfillment center?

2. How many ambulance bases should a city like Ann Arbor, MI have, and where should they be?

3. How many cell phone towers should a mobile phone provider like Verizon have in a city like Ann Arbor, MI, and where should they be?

4. Where should the federal government, through FEMA (the Federal Emergency Management Agency), stockpile emergency supplies for distribution during a natural disaster?

We can divide the literature on facility location modeling into four broad areas.

1. **Analytic models** make many assumptions but are typically easy to solve. Such models provide insights into the solution to more complex problems.

2. **Continuous location models** typically ask where a single facility should be to serve discrete demand locations. This is similar to a center of gravity problem.

3. **Network location models** typically try to find (low-order polynomial) algorithms for location problems on specially structured networks.

4. **Discrete location models** typically result in integer programming problems, which are extensions of linear programming.

We will not cover continuous or network location problems in this introductory text. Readers interested in these topics can refer to other texts including Daskin [2010, 2013]. In this chapter, we will look at a simple analytic model that accounts for *uncertainty* in where demands will occur, *trades off* the fixed facility location costs against the transportation costs, and *optimizes* the tradeoff.

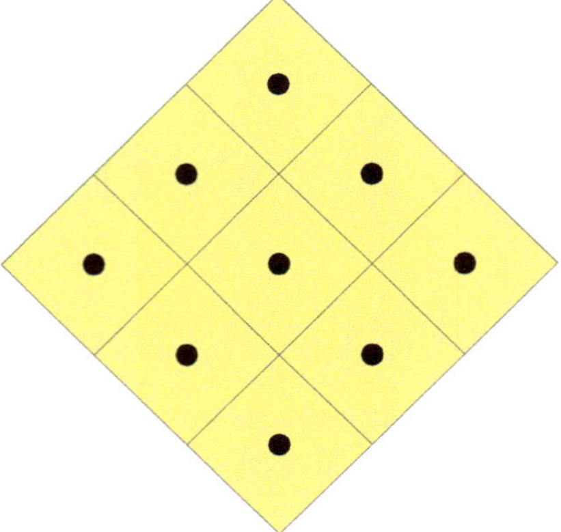

Figure 13.1: Sample service region with nine facilities.

13.2   A SIMPLE ANALYTIC LOCATION MODEL

We will now introduce a simple analytic model to determine how many facilities to have in a region. The model makes many simplifying assumptions but is useful for some of the insights that we can gain from the model.

We begin by assuming that the service region is a square turned 45° with respect to the X- and Y-axes. Figure 13.1 illustrates such a region with nine facilities indicated by blue circles. We will further assume that demands are distributed uniformly over the service region with a density of ρ demands per square mile per year. We can think of this density as the thickness of peanut butter on a piece of bread. It should be intuitively clear that if we had more facilities in the same sized service region—for example if we had 100 facilities in the region shown in Figure 13.1 instead of 9—each demand would, on average, be closer to a facility. The transportation costs would be lower, but the facility construction and operating costs would be higher. The model developed below finds the optimal number of facilities to have.

Next, we assume that area of the square (in square miles) is given by A, so that the length of a side is $\sqrt{A}$. We assume that travel is at 45° to the sides of the square, or parallel to the X- and Y-axes. We assume that the annualized cost of building a facility is f. We assume that it costs c per mile per demand to ship from a facility to a (randomly) located demand. Finally, the key decision variable is N, the number of facilities to locate. WOW! Lots of assumptions!

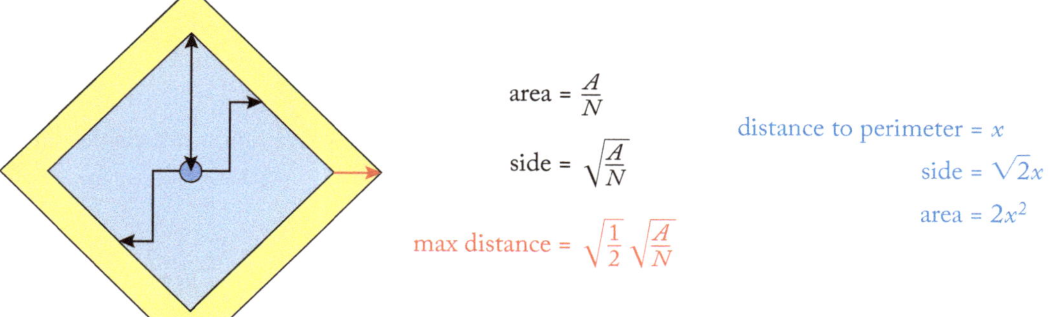

Figure 13.2: A single service region.

The notation is summarized below:

INPUTS

ρ demand density in demands per square mile per year
A area of the square service region
f annualized fixed cost of building and operating a facility
c cost per unit demand per mile of shipping
     from a facility to a demand location

DECISION VARIABLES

N number of facilities to locate

Before we solve for N, we need to compute the expected or average distance between a facility and the demands it serves. In Figure 13.1, each facility would serve demands in the smaller square region in which it is centered. Let us focus on one such region for the moment as shown in Figure 13.2. Since there are N facilities, each facility serves an area of A/N (the yellow shaded area) and the side of such a service area is $\sqrt{A/N}$. The distance between the facility in the center of the region and any point on the perimeter of the region is $\sqrt{A/2N}$. This is the maximum distance of any demand point from a facility.

Now let us consider demands that are within the smaller blue square in Figure 13.2. The distance between the facility and any point on the perimeter of this square is x, so the area of this square is $2x^2$. The distance between the facility and any demand within the blue square is less than or equal to x. So, the probability that the distance is less than or equal to x is simply the area of the blue square, $2x^2$, divided by the area of the entire service region, which is A/N.

In other words, we have

$$P(\text{distance} \leq x) = \frac{2x^2}{A/N}.$$

Now, we also know that the expected value of a non-negative random variable can be computed using $E(X) = \int_0^\infty \{1 - F_X(x)\}dx$. In our case, we have

$$E(\text{distance}) = \int_0^{\sqrt{A/2N}} \left\{1 - \frac{2x^2}{A/N}\right\} dx = \frac{2}{3}\sqrt{\frac{A}{2N}}.$$

In other words, the expected distance between a facility in the middle of a square (yellow) region of size A/N is $\frac{2}{3}\sqrt{\frac{A}{2N}}$ or two-thirds of the maximum distance between the facility and the perimeter of the region.

We can now write down the total cost as a function of N, the number of facilities in the square region of size A, the big yellow square of Figure 13.1. This is given by

$$\text{Total cost}(N) = \underbrace{fN}_{\substack{\text{fixed} \\ \text{cost per year}}} + \underbrace{\left(\frac{2}{3}\sqrt{\frac{A}{2N}}\right)}_{\text{avg. dist.}} \cdot \underbrace{c}_{\$/\text{dem-mile}} \cdot \underbrace{\rho A}_{\substack{\text{demands} \\ \text{per year}}}.$$

The facility cost grows linearly with the number of facilities as shown by the first term, fN. The second term is the product of the average distance between a demand and the facility that serves it, the cost per demand-mile, and the number of demands per year. The second term gives the expected transportation cost, while the first term gives the facility cost. Figure 13.3 plots the facility cost (in blue), the transport or mileage cost (in purple), and the total cost (in red).

If we now take the derivative of this function with respect to the number of facilities, N, and set that equal to zero, we get

$$N^* = A\left(\frac{c\rho\sqrt{2}}{6f}\right)^{2/3}.$$

The optimal number of facilities, N^*, goes up linearly with the area served, and goes up as the 2/3 power of the unit shipping cost, c, and the density of demand, ρ; and it goes down with the 2/3 power of the fixed facility costs, f. The directions of these relationships make sense. For example, if we double the service area—if a company goes from serving only the eastern half of the United States to serving the entire contiguous 48 states—the number of facilities it uses should double.

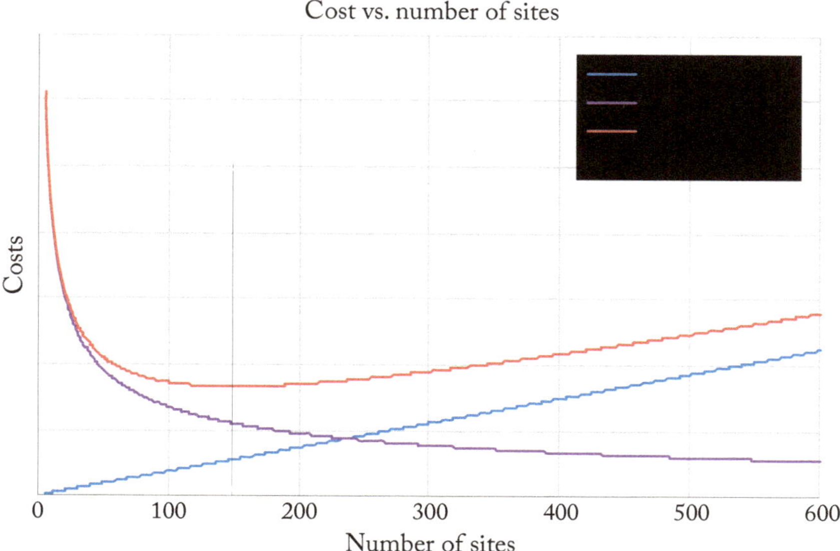

Figure 13.3: Fixed and mileage or transport cost and total cost as a function of the number of facilities.

We can now substitute this optimal number of facilities into the total cost function to obtain the optimal total cost as follows:

$$TC^*(N^*) = Af^{1/3}(c\rho)^{2/3}\left[\underbrace{\left(\frac{\sqrt{2}}{6}\right)^{2/3}}_{\text{fixed}} + \underbrace{2\left(\frac{\sqrt{2}}{6}\right)^{2/3}}_{\text{transport}}\right]$$

$$\approx 1.1447 \cdot Af^{1/3}(c\rho)^{2/3}.$$

Note that at the optimal number of facilities, the facility or fixed costs should be about half the transport costs. Also note that, unlike the EOQ model whose graph looks similar to Figure 13.3, the optimal value does *not* occur where the facility and transport curves intersect. We need to take the derivative of the total cost function and set it equal to zero.

This model makes **many** simplifying assumptions as noted above. One additional slight of hand that we made is that we treated the number of facilities deployed, N, as if it were a continuous real number so that we could then take the derivative of the total cost function with respect to this value. In fact, the number of facilities needs to be an integer. Furthermore, in deriving the average distance, we assumed that we could divide the full service region into equally sized square regions, each of size A/N. Clearly, this is only true if N itself is the square

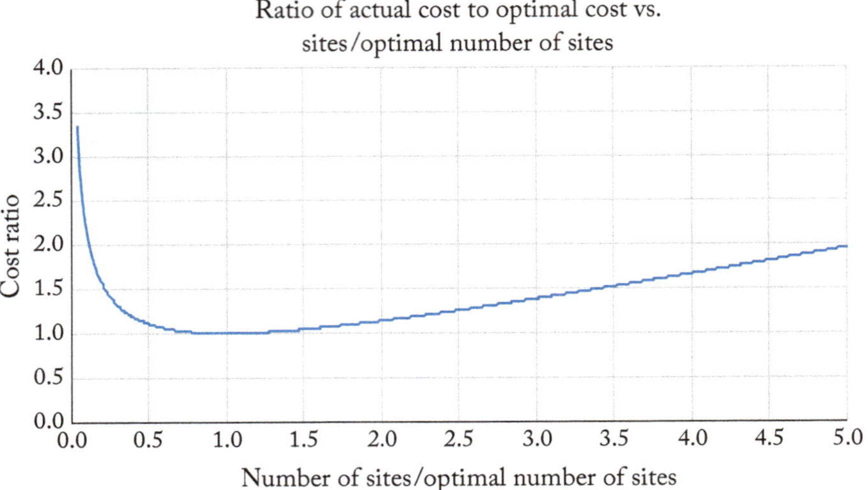

Figure 13.4: Ratio of actual cost to optimal cost as a function of the ratio of the number of used facilities to the optimal number of facilities.

of an integer (or a number like 1, 4, 9, 16, ...). If we had two facilities, for example, we could not divide the big square into two equally sized squares of area $A/2$. In the next two sections, we examine these issues in some more detail.

13.3   SENSITIVITY OF THE SOLUTION TO CHANGES IN THE NUMBER OF FACILITIES USED

The number of facilities given by the equation for N^* could easily be a fraction. In fact, for $A = 3{,}120{,}000$, $\rho = 105$, $c = 0.01$, and $f = 750{,}000$, we get $N^* = 148.987$ and a total cost of $TC(N^*) = \$335{,}221{,}322$. These values roughly approximate the area and density of the contiguous United States. They are also the values used in plotting Figure 13.3. Clearly, we cannot locate 148.987 facilities. Therefore, we may want to know how far off the total cost will be from the optimal cost if, instead of using N^* facilities, we use βN^*, where β is any positive number. For example, if $\beta = 1.5$ we would be using 50% more facilities than the optimal number. We can show that

$$\frac{TC(N)}{TC(N^*)} = \frac{\beta + 2/\sqrt{\beta}}{3}.$$

Figure 13.4 plots this ratio as a function of β. If the number of facilities is between 50% of the optimal number and 180% of the optimal number, the total cost remains within 10% of the optimal value. The total cost is relatively insensitive to the number of facilities we use. This is good news. For the example above, this ratio is 1.000288 if we locate 144 (or 12^2) facilities.

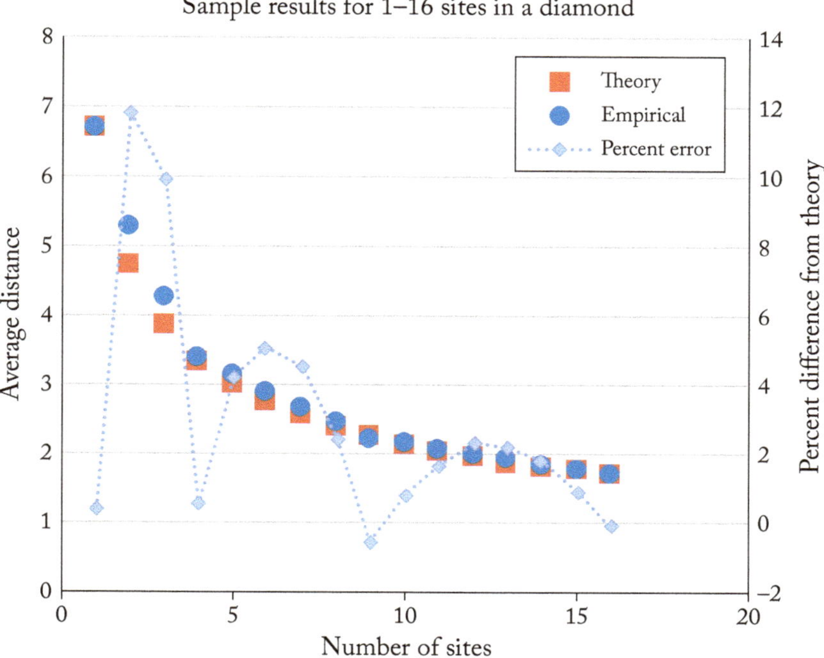

Figure 13.5: Theoretical distance, actual average distance, and percent difference.

This means that the total cost would be $96.6 thousand more than the optimal total cost of over $335 million. This is a tiny amount relative to the total annual cost.

13.4   INACCURACIES DUE TO OUR INABILITY TO DIVIDE THE SERVICE REGION EVENLY

In this section, we examine the impact of our inability to divide the big service region evenly into N smaller service regions on the average distance. To do this, we simulated (or sampled) 12,000 demand points in a square service area and then found the optimal (or near optimal) locations for N facilities and computed the average distance between the 12,000 demand points and the nearest facility. The distance from the center of the region to one of the corners of the square service area was 10 and so the service region had an area of 200 square units.

Figure 13.5 shows the theoretical distance, $\frac{20}{3\sqrt{N}}$, and the actual average or empirical distance. The figure also shows the percent error along the right-hand Y-axis. The percent error due to our inability to divide the region into N smaller squares is small, particularly once the number of facilities being located exceeds 9.

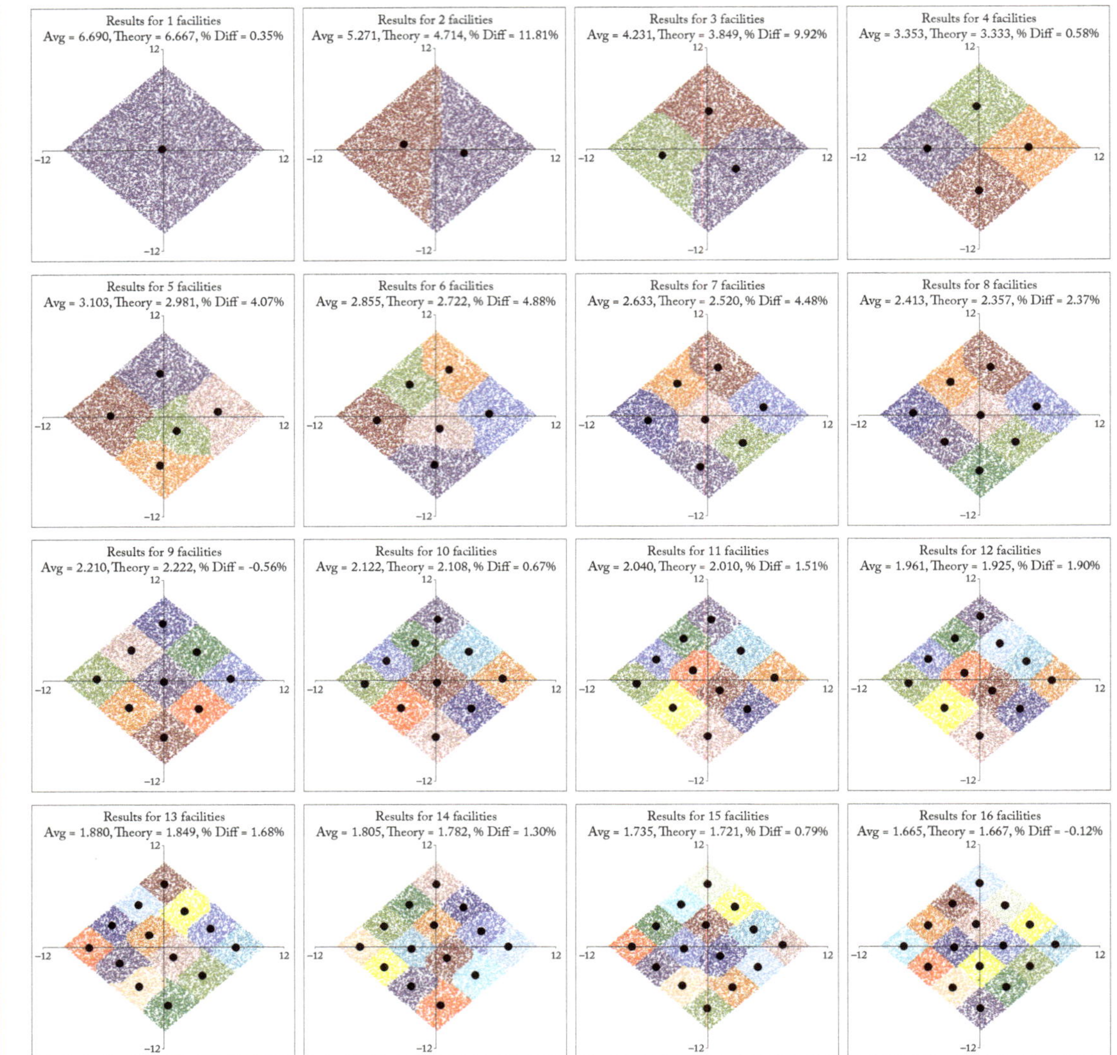

Figure 13.6: Near optimal service regions for $N - 1, 2, \ldots, 16$ facilities.

Finally, Figure 13.6 shows the simulated results. For 1, 4, 9, and 16 facilities, the big service region is divided (approximately) into an appropriate number of smaller square sub-regions with the facilities in the center of the sub-regions, as expected. For other values of the number of facilities, the sub-regions are not square. For example, with only two facilities, the sub-regions are roughly triangular, resulting in the empirical distance being about 12% above the theoretical value.

13.5   SUMMARY

This chapter has shown how we can, with enough simplifying assumptions, develop a model that can explore the tradeoff between paying for more facilities and reducing the transportation cost, or having fewer facilities but being further from the demands. It is worth remembering that the model made many simplifying assumptions. For example, we assumed that the demand was uniformly distributed across the service region. If we think about the United States, this is anything but true. According to Wikipedia, if you apply the population density of Alaska to the island of Manhattan, you would have about 29 people on the island. Many apartment buildings in the city have more people than that. If you applied the population density of Manhattan to Alaska, you would have about 40.8 billion people in Alaska, or roughly 5.25 times the number of people who inhabit the earth! The equal demand density assumption is perhaps the most egregious assumption that we made in the development of this model. In the next two chapters, we explore two different discrete location models that allow us to relax this very strong assumption along with other assumptions.

13.6   REFERENCES

M. S. Daskin, *Service Science*, John Wiley & Sons, Inc., New York, 2010. DOI: 10.1002/9780470877876. 87

M. S. Daskin, *Network and Discrete Location: Models, Algorithms and Applications*, 2nd ed., John Wiley & Sons, Inc., New York, 2013. DOI: 10.1002/9781118537015. 87

CHAPTER 14

Maximizing Demand Coverage

14.1 DISCRETE LOCATION MODELS

One of the many problems with the simple analytic location model of Chapter 13 was that it assumed that demand was uniformly distributed. But the population is anything but uniformly distributed across the United States, for example. The density of New Jersey, with over 1,200 people per square mile, is 200 times the density of Wyoming, which has only 6 people per square mile. For this reason, analytic location models can provide qualitative insights into the structure of location problems, but are rarely used for actual decision-making. In this chapter, we introduce a simple discrete location model that has been used extensively in locating emergency service facilities (fire stations and ambulance bases) and in many other problem domains.

In discrete location models, we divide the region under study into small sub-regions and represent each sub-region by a node. Sub-regions might correspond to Census tracts, or Census block groups, or counties, or zip codes, or any other level of aggregation for which the data are available. There is generally a tradeoff between using too few sub-regions and thereby not being able to have a high fidelity model and using too many sub-regions and losing computational tractability, since the solution time for most algorithms increases significantly with the size of the problem being modeled. Generally, we also assume that the set of candidate facility sites is the same as the set of demand nodes, though this is not a needed assumption.

In this chapter and the next two chapters, we will use the 33-node Census tract representation of Ann Arbor, MI. Figure 14.1 shows the Census tract population in Ann Arbor by quartiles. The tract populations (according to the 2010 Census) range from 1,551–7,098. The total population in the dataset is 115,103.

Suppose we want to locate ambulance bases in Ann Arbor to maximize the number of people who are within 1.5 miles of the nearest base. Such individuals are said to be *covered* by an ambulance base, while those who live further away from the nearest base are said to be *uncovered*. It is important to note that *all* individuals would be served by the ambulance service, whether they are covered or not. The notion of *coverage*, however, is a measure of the quality of the service. In some sense, the more people who are covered, the better the service. Since we have aggregated the demand into Census tracts that are represented by nodes, if a node is within 1.5 miles of an ambulance base, the node will be covered and all the demand in the tract will be counted as covered.

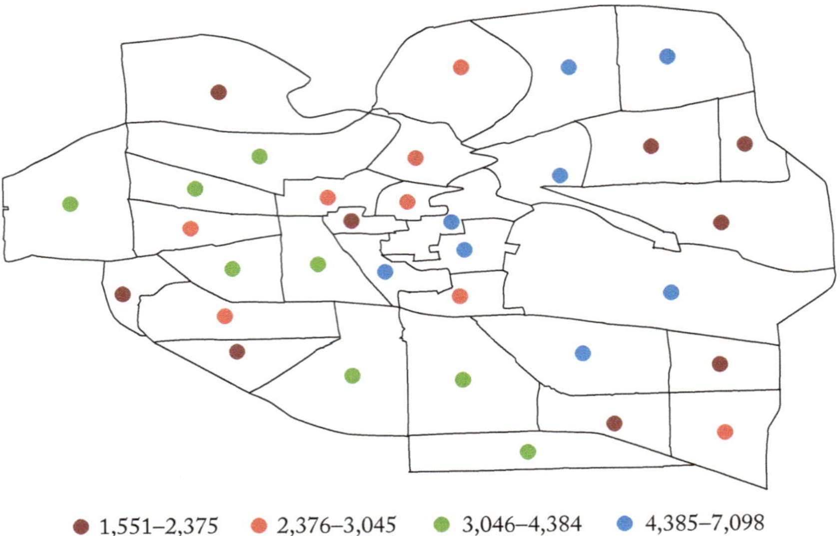

Figure 14.1: Population distribution of Ann Arbor.

14.2   THE MAXIMAL COVERING LOCATION MODEL

The model we formulate in this chapter maximizes the number of covered individuals. This is the maximal covering model [Church and ReVelle, 1974]. To do so, we define the following inputs and sets:

INPUTS

I	set of demand nodes
J	set of candidate locations
h_i	demand at node $i \in I$
d_{ij}	distance between candidate location $j \in J$ and demand node $i \in I$
d^c	critical coverage distance
a_{ij}	$\begin{cases} 1 & \text{if } d_{ij} \leq d^c \\ 0 & \text{if not} \end{cases}$
p	number of facilities to locate

The demand at a node will be taken to be the population. This does not mean that everyone in a Census tract will need ambulance service every year or every day, but rather that the demand for ambulance service is (roughly) proportional to the population of a tract. For the distances between nodes, we will be using the great circle distances (the straight line distances on a sphere), which are easy to compute given the longitude and latitude of each node. The rest of the notation is self-explanatory.

The key decisions are where to locate facilities or, in our case, ambulance bases. The second decision variable, Z_i, will allow us to compute the total covered demand. This notation is summarized below:

DECISION VARIABLES

$$X_j \quad \begin{cases} 1 & \text{if we locate at candidate site } j \in J \\ 0 & \text{if not} \end{cases}$$

$$Z_i \quad \begin{cases} 1 & \text{if we demand node } i \in I \text{ is covered} \\ 0 & \text{if not.} \end{cases}$$

With this notation, the model can be formulated as follows:

$$\begin{array}{llll} Max & \displaystyle\sum_{i \in I} h_i Z_i & & \text{total covered demand} \\[2ex] s.t. & \displaystyle Z_i - \sum_{j \in J} a_{ij} X_j \leq 0 & \forall i \in I & \text{linkage constraint} \\[2ex] & \displaystyle\sum_{j \in J} X_j = p & & \text{locate } p \text{ facilities} \\[2ex] & X_j \in \{0, 1\} & \forall j \in J & \text{integrality} \\[1ex] & 0 \leq Z_i \leq 1 & \forall i \in I & \text{bounds on coverage variables.} \end{array}$$

$h_i Z_i$ will be 0 if demand node i is not covered and will equal the demand at node i, h_i, if the node is covered. The objective function is then equal to the total covered demand. The first constraint says that a node cannot be counted as covered unless we locate at least one facility within the coverage distance of the node. Specifically, the summation in the first constraint, $\sum_{j \in J} a_{ij} X_j$, counts the total number of located facilities that are within the coverage distance of node i. If this total is 1 or more, then node i can be counted as being covered ($Z_i = 1$); otherwise, the node must be uncovered ($Z_i = 0$). The second constraint says that we are to locate p facilities. Next we force the location variables to be binary. Finally, we have bounds on the coverage variables, Z_i. Note that these variables will naturally be either 0 or 1 and so we do not explicitly need to state that they have to be binary.

Figure 14.2 gives the optimal solution to this problem for the 33-node Ann Arbor dataset with $p = 5$ and $d^c = 1.5$. Roughly 112,600 of the 115,000 people are covered, representing almost 98% of the total population. There is one tract in the southeastern part of the city that is not covered.

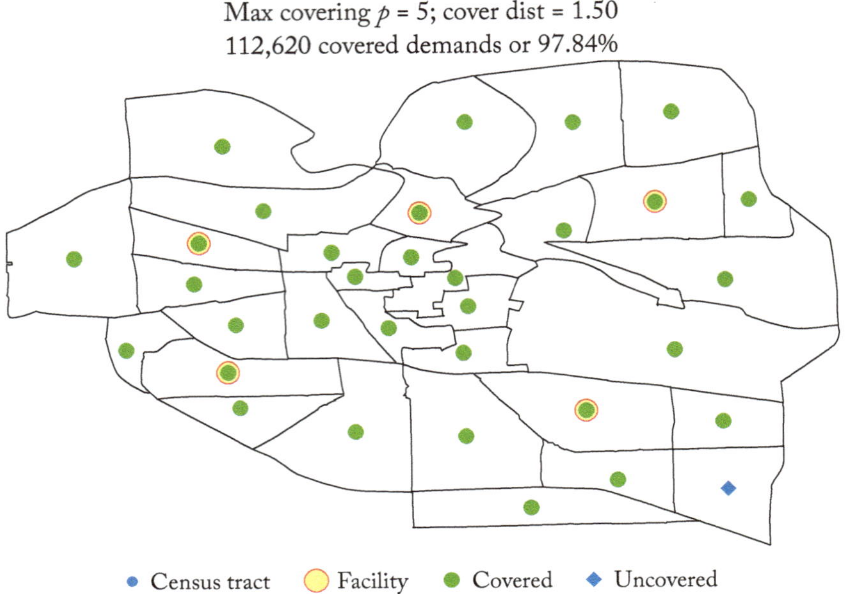

Figure 14.2: Optimal solution to the maximal covering problem with a coverage distance of 1.5 miles and 5 facilities in Ann Arbor, MI.

Figure 14.3 plots the percent of the demand that is covered vs. the number of facilities that are located for three different coverage distances: 1.0 mile, 1.25 miles, and 1.5 miles. For any coverage distance, as we increase the number of facilities that we locate the percent of the total demand or population that is covered goes up, but generally at a decreasing rate. Also, for any fixed number of facilities, increasing the coverage distance increases the percent of the demand that is covered (or leaves it the same).

Another problem of interest is the *set covering* problem which finds the minimum number of facilities needed to cover all demands. Figure 14.3 allows us to find the solution to the set covering problem for the Ann Arbor dataset. All we need to do is to fix the coverage distance and then find the solution at which 100% of the demand is covered. When the coverage distance is 1.0 mile, we need 12 facilities to cover all demands; when the distance is 1.25 miles, only 8 facilities are needed; and when the coverage distance is 1.5 miles, 6 facilities are needed.

Figure 14.3 illustrates a general rule of thumb for the maximal covering problem. If we locate half of the number of facilities needed to cover all demands within a given coverage distance, we generally can cover 80% or more of the total demand. The last half of the facilities that are located cover only 20% of the demand, clearly illustrating the decreasing marginal returns to scale associated with adding extra facilities. If we solve the maximal covering problem using the 3,109 counties in the contiguous United States, we need 7 facilities to cover all demand nodes

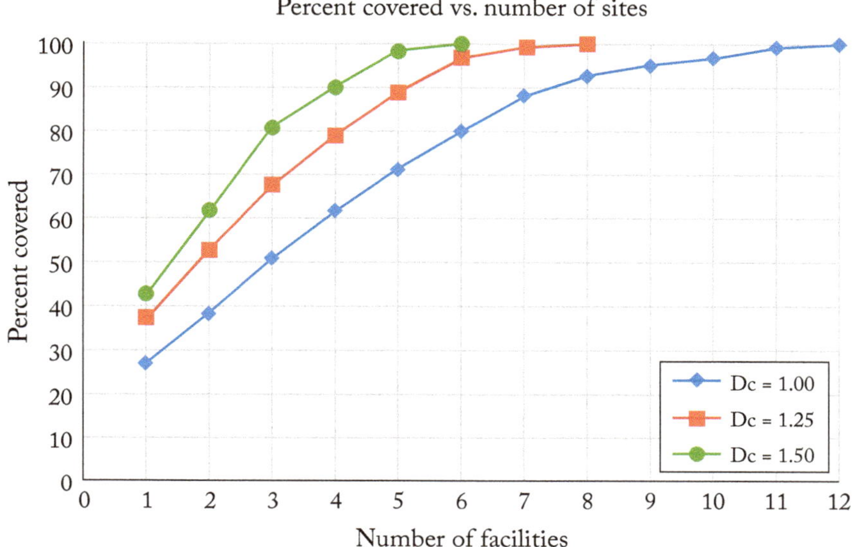

Figure 14.3: Impact of the number of facilities and the coverage distance on the percent of the population covered.

with a coverage distance of 500 miles. With three facilities we can cover over 82% of the total demand and with 4 facilities we can cover over 90% of the total demand. This illustrates that the rule of thumb works for much larger-scale problems as well.

14.3   REFERENCES

R. Church and C. ReVelle, The maximal covering location problem, *Papers of the Regional Science Association*, 22:101–118, 1974. DOI: 10.1111/j.1435-5597.1974.tb00902.x. 98

CHAPTER 15

Minimizing the Average Distance to a Facility

15.1 DEMAND-WEIGHTED AVERAGE DISTANCE

In Chapter 14, we introduced a simple discrete location model that maximizes the number of covered demands. This sort of model is often used in locating public service facilities. For commercial facilities, like warehouses and fulfillment centers, it is often more appropriate to minimize the *demand-weighted average distance* between customers or demand nodes and the nearest facility. This is the focus of this chapter.

We begin by discussing what the demand-weighted average distance means. Consider the simple network shown in Figure 15.1 with three nodes. The distances between the nodes are shown beside the links and the demand at each node is shown beside each node with blue lettering. The total demand in the network is 1,000 units.

Table 15.1 demonstrates the computation of the demand-weighted total and average distance for the network of Figure 15.1 if we locate a single facility at each of the candidate nodes. For example, if we locate at node A, then node A contributes 0 to the demand-weighted total distance, since the distance from node A to node A is 0. Node B, on the other hand, contributes 1,350 (3 times 450) to the demand-weighted total distance and node C contributes 2,000 (5 times 400) to the demand-weighted total distance. The demand-weighted total distance is 3,350 if we locate at node A. Since the total demand is 1,000, we can simply divide the demand-weighted total distance by 1,000 to get the demand-weighted average distance of 3.35 if we locate a single facility at node A. Similarly, if we locate a single facility at node B or node C, the demand-weighted average distance is 2.05 and 2.55, respectively.

15.2 THE P-MEDIAN PROBLEM

Now that we understand the notion of the demand-weighted total or average distance, we can formulate an integer linear programming problem to identify the locations of p facilities to minimize the demand-weighted total distance. To do so, we define the following inputs and decision variables:

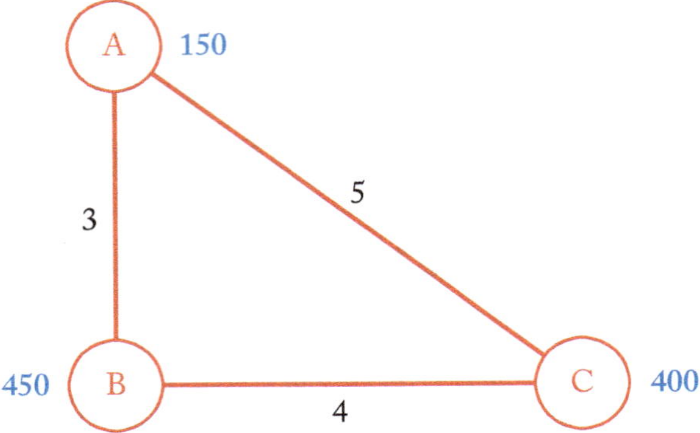

Figure 15.1: Simple network to illustrate demand-weighted average distance.

Table 15.1: Computation of the demand-weighted total and average distance for the network of Figure 15.1

Node	Locate at		
	A	B	C
A	0	450	750
B	1,350	0	1,800
C	2,000	1,600	0

	A	B	C
Demand-weighted total	3,350	2,050	2,550
Demand-weighted average	3.35	2.05	2.55

INPUTS

I set of demand nodes

J set of candidate locations

h_i demand at node $i \in I$

d_{ij} distance between candidate location $j \in J$ and demand node $i \in I$

p number of facilities to locate

DECISION VARIABLES

$$X_j \quad \begin{cases} 1 & \text{if we locate at candidate site } j \in J \\ 0 & \text{if not} \end{cases}$$

$$Y_{ij} \quad \begin{cases} 1 & \text{if we demand node } i \in I \text{ is assigned to a facility at candidate site } j \in J \\ 0 & \text{if not.} \end{cases}$$

The inputs are basically the same as those used for the maximal covering problem of Chapter 14. Also, the location variable, X_j, is the same as that used in Chapter 14. Now, however, instead of a coverage variable, Z_i, we need an assignment variable, Y_{ij}, which will be 1 if the demand at node i is assigned to a facility at node j and 0 if it is not assigned to a facility at node j. With this notation, we can now formulate the p-median problem as follows:

$$
\begin{aligned}
\text{Min} \quad & \sum_{i \in I} \sum_{j \in J} h_i d_{ij} Y_{ij} && \text{demand-weighted total distance} \\
\text{s.t.} \quad & \sum_{j \in J} Y_{ij} = 1 && \forall i \in I && \text{each node is assigned} \\
& Y_{ij} - X_j \leq 0 && \forall i \in I, j \in J && \text{assign only to open facilities} \\
& \sum_{j \in J} X_j = p && && \text{locate } p \text{ facilities} \\
& X_j \in \{0, 1\} && \forall j \in J && \text{integrality} \\
& Y_{ij} \geq 0 && \forall i \in I, j \in J && \text{non-negative variables.}
\end{aligned}
$$

$h_i d_{ij}$ represents the demand at node i multiplied by the distance from i to j. This would be the contribution of node i to the demand-weighted total distance if node i is assigned to a facility at node j. We multiply this by Y_{ij} and then sum over all candidate sites j, to obtain the actual contribution of node i to the demand-weighted total distance. This is then summed over all demand nodes to get the demand-weighted total distance in the network. The first constraint ensures that each demand node is assigned to exactly one facility. The second constraint stipulates that a demand node cannot be assigned to facility unless we actually locate a facility at that site, or $X_j = 1$. The third constraint says that we must locate exactly p facilities. Finally, constraint 4 says that the location variables must be binary and constraint 5 says that the assignment variables must be non-negative. Note that the assignment variables will always take on a value of either 0 or 1, unless there are ties in the distance matrix. If an assignment variable is fractional, because of a tie in the distance matrix, we can increase one of the values for that demand node to 1 and set all the other fractional values for that demand node to 0, without changing the value of the objective function. The p-median problem dates back to the seminal work of Hakimi [1964, 1965].

Figure 15.2 shows the 5-median solution for the 33-node Ann Arbor dataset. The demand-weighted total distance is nearly 90,000 units. The demand-weighted average distance

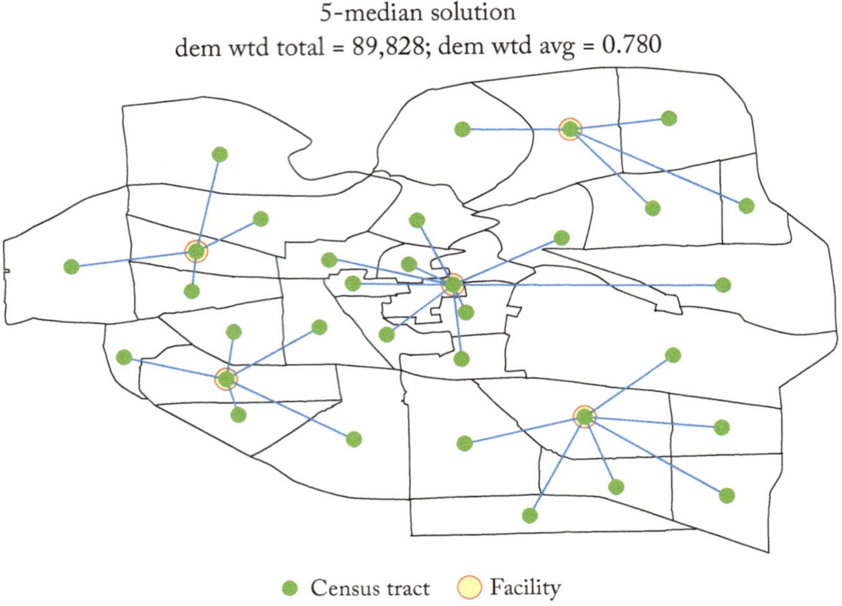

Figure 15.2: 5-Median solution for the 33-node Ann Arbor Dataset.

is 0.78 miles, a number that is much easier to interpret and understand than is the demand-weighted total distance. The five facilities are shown as red circles. Nodes of Census tracts in which a facility is not located are connected to the nearest facility by a thin blue line. These form what is called a spider network.

Figure 15.3 shows the relationship between the demand-weighted average distance and the number of facilities that are sited. As the number of facilities increases, the demand-weighted average decreases, though (generally) at a decreasing rate, again suggesting that there are decreasing marginal benefits associated with adding facilities in the p-median problem.

15.3   THE UNCAPACITATED FIXED CHARGE LOCATION PROBLEM–AN EXTENSION OF THE P-MEDIAN PROBLEM

While Figure 15.3 shows the relationship between the number of facilities and the demand-weighted average distance, if we know the annual fixed cost of locating at each candidate location, f_j, and the transportation cost per unit per mile, α, we can minimize the total annual cost of the facilities plus the transportation cost, assuming that the demands are given as annual demands. The revised formulation of the *uncapacitated fixed charge location* problem becomes:

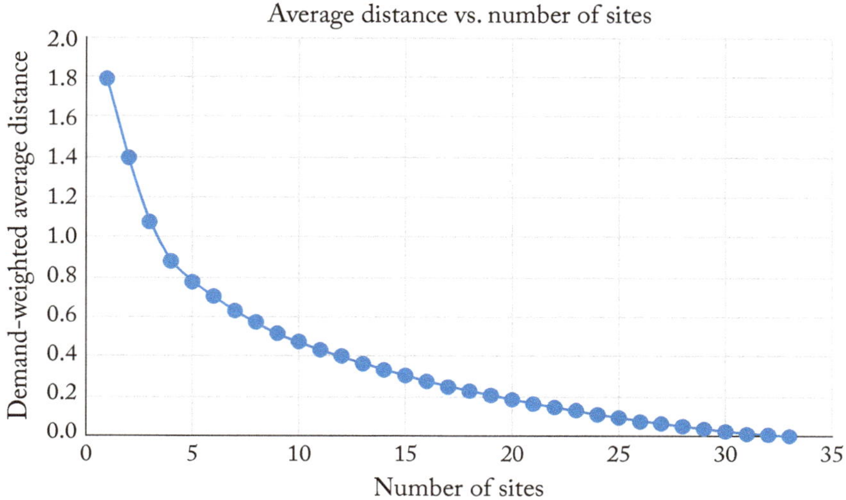

Figure 15.3: Average distance vs. number of facilities for the 33-node Ann Arbor Dataset.

$$
\begin{aligned}
&\textit{Min} \quad \sum_{j \in J} f_j X_j + \alpha \sum_{i \in I} \sum_{j \in J} h_i d_{ij} Y_{ij} && \text{facility plus transport costs} \\
&\textit{s.t.} \quad \sum_{j \in J} Y_{ij} = 1 && \forall i \in I && \text{each node is assigned} \\
&\qquad Y_{ij} - X_j \le 0 && \forall i \in I;\ j \in J && \text{assign only to open facilities} \\
&\qquad X_j \in \{0, 1\} && \forall j \in J && \text{integrality} \\
&\qquad Y_{ij} \ge 0 && \forall i \in I;\ j \in J && \text{non-negative variables.}
\end{aligned}
$$

The constraints remain the same, except that we now must eliminate the third constraint in the p-median formulation of Section 15.2 that stipulates that we locate p facilities. As we increase α, the transportation cost per item per mile, we will eventually want to add another facility. This happens when the incremental facility cost is less than the decrease in the transportation cost that would result from adding a facility, thereby reducing the second term of the objective function above.

If we set the annualized fixed cost of locating at any location equal to $1,000,000, Figure 15.4 plots the results of increasing the cost per item per mile from $0 to $100. As we increase α, we progressively add facilities as shown by the red facility cost curve. For a fixed number of facilities (where the red line is flat), as we increase α, the total demand-weighted distance remains constant (as shown by the black line) but the transportation *cost* increases linearly due to the increase in α. Finally, the blue line shows the sum of the facility and transport costs. It is a piecewise linear function with a decreasing slope as we increase the cost per item per mile.

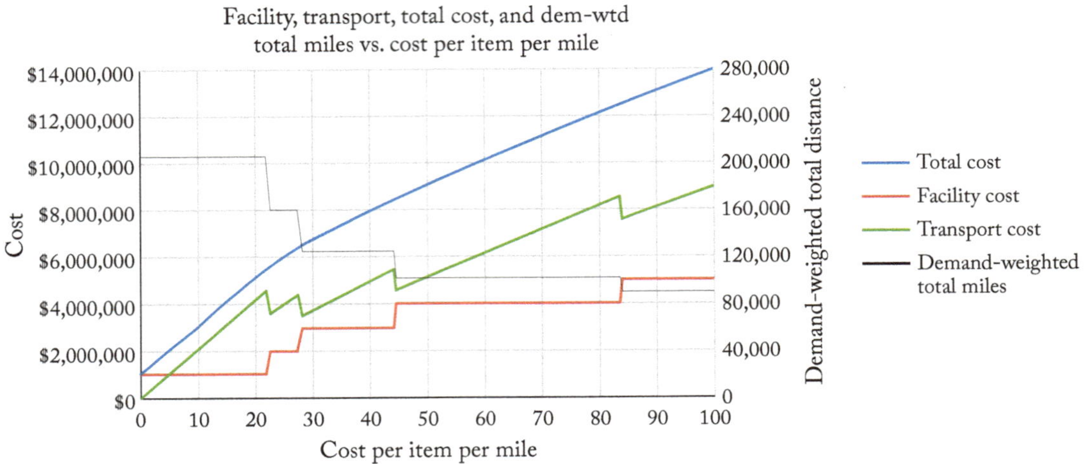

Figure 15.4: Total cost, transport cost, facility cost, and demand-weighted total miles vs. cost per item per mile.

The total cost curve in Figure 15.4 is actually the lower envelope of the cost curves associated with locating each number of facilities. This is shown in Figure 15.5, which plots the total facility and transport cost if we locate one through five facilities. Each curve is a straight line; the slope of the lines decreases as we increase the number of facilities. Thus, the light blue line shows the total cost as a function of the cost per item per mile if we locate only a single facility. The red line shows the total cost if we locate two facilities, and so on. The darker blue line is again the total cost as a function of the cost per item per mile. It is the lower envelope of the five cost lines shown in the graph. Finally, the black step-function in the graph shows the optimal number of facilities as a function of the cost per item per mile.

Figure 15.6 plots the facility (red), transport (green), and total (blue) cost as a function of the cost per item per mile, α, with this cost going from \$0 to \$1,500. Above a cost per item per mile of \$1,389.75, it is optimal to locate at every node or Census tract, thereby eliminating the transport cost. The total cost in this case hits a maximum of \$33,000,000 (or 33 facilities times \$1,000,000 per facility).

Finally, the reader should note that this model is the discrete analog of the analytic location model discussed in Chapter 13. There are three key differences. First, this model accounts for different demand levels across the region, while the model of Chapter 13 assumed that demand was uniformly distributed. Second, the facility costs can differ for each candidate location in this model (though we set them all equal to \$1,000,000 in this example), while the facility costs in Chapter 13 were all equal. Third, the total cost has a maximum which is just the sum of the fixed facility costs at all locations in this model, or \$33,000,000 in this case, while the total cost in Chapter 13 grew as the 2/3 power of the cost per item per mile.

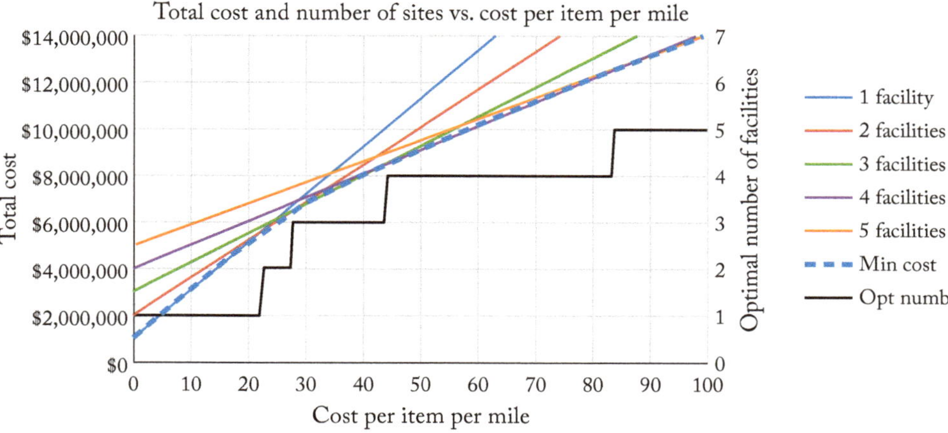

Figure 15.5: Total cost and number of sites vs. cost per item per mile.

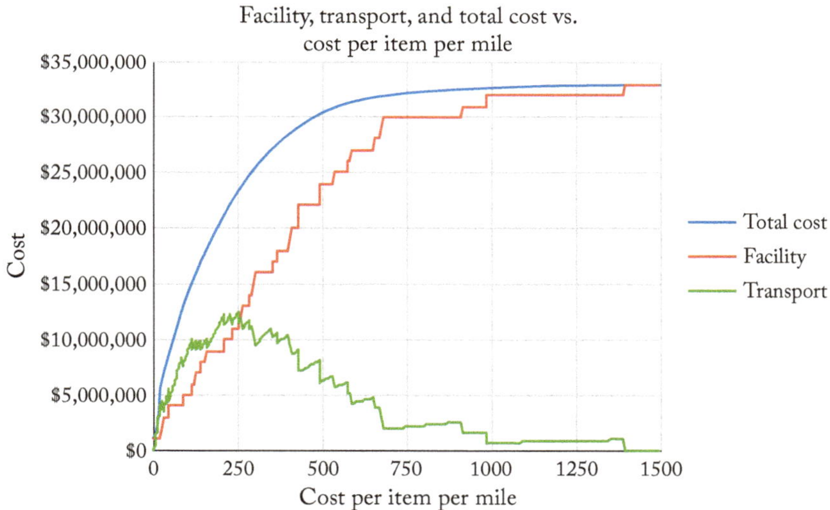

Figure 15.6: Total cost, facility cost, and transport cost vs. cost per item per mile.

The next chapter will explore the tradeoff between covering demand and the demand-weighted average distance. This will emphasize the importance of *tradeoffs* in operations management in general and location modeling in particular.

15.4   REFERENCES

S. L. Hakimi, Optimum location of switching centers and the absolute centers and medians of a graph, *Operations Research*, 12:450–459, 1964. DOI: 10.1287/opre.12.3.450. 105

S. L. Hakimi, Optimum distribution of switching centers in a communication network and some related graph theoretic problems, *Operations Research*, 13:462–475, 1965. DOI: 10.1287/opre.13.3.462. 105

CHAPTER 16

A Facility Location Tradeoff

16.1 A NEW KIND OF TRADEOFF

One of the key themes of operations management, we have argued, is the analysis of trade-offs. The need to make tradeoffs permeates our personal, professional, and public lives. We have already seen tradeoffs between the number of facilities we deploy and the percent of the population that is covered (in Chapter 14) and the demand-weighted average distance (in Chapter 15). These are both examples of the tradeoff between the achievement of a single objective (be it coverage or average distance) and the resources devoted to achieving that objective.

In the economic order quantity model of Chapter 8, the newsvendor model of Chapter 9, and the analytic location model of Chapter 13, we were trading off two types of costs. For the EOQ model, for example, we were trading off the fixed costs of placing orders and the inventory carrying costs. In the newsvendor problem, we traded off the cost of losing money on unsold items against the loss of revenue that resulted from having too few items to sell. Finally, in the analytic location model, we examined the tradeoff between the facility costs and the expected transportation costs to (or from) the facilities from (or to) the customers.

In many contexts, particularly in the public sector, we need to make tradeoffs between competing objectives for the same level of resource deployment. In this chapter we will explore one such tradeoff: the tradeoff between minimizing the demand-weighted average distance to the nearest facility on the one hand and maximizing the percent of the demand that is covered on the other hand.

To motivate this analysis, consider the 33-node Census tract representation of Ann Arbor. Figure 16.1 shows the solution to the 6-median solution. The demand-weighted average distance is 0.703 miles. 32.56% of the demands are within 0.5 miles of a facility. Covered nodes are shown in green. Figure 16.2 shows the solution to the maximal covering problem. 36.6% of the demand is within 0.5 miles of a facility, representing a 12.4% improvement in coverage over the 6-median solution shown in Figure 16.1. However, this comes at the price of a 25.1% degradation in the average distance which increases from 0.703 miles in the 6-median solution of Figure 16.1 to 0.879 miles in the maximal covering solution of Figure 16.2.

The key questions we will ask and answer in this chapter are (1) are there good compromise solutions between these two extremes and (2) how can we find such solutions. The short answer to the first question is that there are almost always good compromise solutions between solutions that optimize single objectives. Thus, our focus will be on how to find these solutions. We will illustrate one of the techniques—the constraint method—by exploring the tradeoff between the

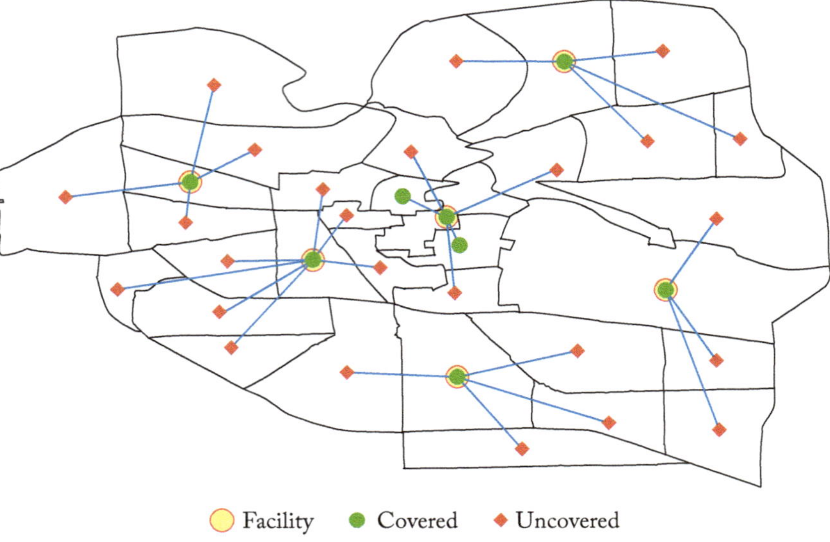

Figure 16.1: 6-Median solution with a coverage distance of 0.5 miles.

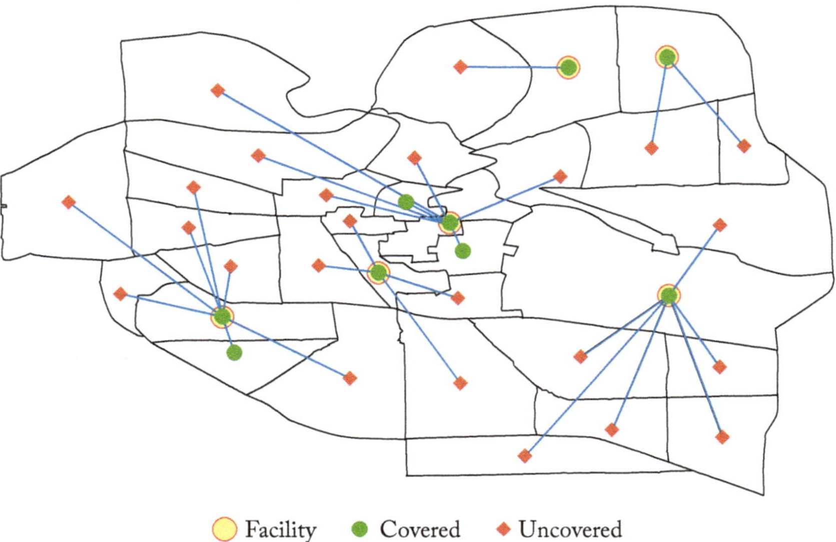

Figure 16.2: Maximal covering solution with 6 facilities and a coverage distance of 0.5 miles.

median and covering objectives. The same technique, appropriately modified, can be used for other tradeoff analyses.

The goal of a *multi-objective* analysis is to find the **set** of *non-dominated* solutions. **Solution A *dominates* solution B**, if each of the objective function values associated with solution A are at least as good as the corresponding objective function values of solution B, and, for at least one objective function value, solution A is strictly better than solution B. Neither of the solutions shown in Figures 16.1 and 16.2 dominates the other.

16.2   THE MEDIAN-COVERING TRADEOFF

To formulate the problem of finding the tradeoff between the p-median problem and the maximal covering problem, let us define the following inputs:

INPUTS

I	set of demand nodes
J	set of candidate locations
h_i	demand at node $i \in I$
d_{ij}	distance between candidate location $j \in J$ and demand node $i \in I$
p	number of facilities to locate
d^c	critical coverage distance
a_{ij}	$\begin{cases} 1 & \text{if } d_{ij} \leq d^c \\ 0 & \text{if not} \end{cases}$
w	weight on demand-weighted total distance; $0 \leq w \leq 1$
$h^c_{\min}$	minimum demand that must be covered

These inputs are identical to those of the maximal covering problem (Chapter 14) and the p-median problem (Chapter 15). In addition to those inputs, we have two new inputs shown in blue. They will be discussed below. In addition, we define the three classes of decision variables used in those two models as follows:

DECISION VARIABLES

X_j	$\begin{cases} 1 & \text{if we locate at candidate site } j \in J \\ 0 & \text{if not} \end{cases}$
Y_{ij}	$\begin{cases} 1 & \text{if we demand node } i \in I \text{ is assigned to a facility at candidate site } j \in J \\ 0 & \text{if not} \end{cases}$

$$Z_i \begin{cases} 1 & \text{if we demand node } i \in I \text{ is covered} \\ 0 & \text{if not} \end{cases}$$

We can now formulate a multi-objective model that minimizes the demand-weighted total distance (the p-median objective) and maximizes the covered demand as shown below:

$$
\begin{array}{lll}
\textit{Min} & \sum_{i \in I} \sum_{j \in J} h_i d_{ij} Y_{ij} & \text{demand-weighted total distance} \\
\textit{Max} & \sum_{i \in I} h_i Z_i & \text{total covered demand} \\
\textit{s.t.} & \sum_{j \in J} Y_{ij} = 1 \quad \forall i \in I & \text{each node is assigned} \\
& Y_{ij} - X_j \leq 0 \quad \forall i \in I, j \in J & \text{assign only to open facilities} \\
& \sum_{j \in J} X_j = p & \text{locate } p \text{ facilities} \\
& Z_i - \sum_{j \in J} a_{ij} X_j \leq 0 \quad \forall i \in I & \text{linkage constraint} \\
& X_j \in \{0, 1\} \quad \forall j \in J & \text{integrality} \\
& Y_{ij} \geq 0 \quad \forall i \in I, j \in J & \text{non-negative variables} \\
& 0 \leq Z_i \leq 1 \quad \forall i \in I & \text{bounds on coverage variables}
\end{array}
$$

The first objective and the first three constraints are identical to those of the p-median problem, while the second objective and the fourth constraint are identical to the maximal covering problem. The last three sets of constraints define the feasible values of the location, allocation, and coverage variables.

Unfortunately, optimization algorithms allow us to use only a single objective function, not two or more functions. To deal with this problem, we can either optimize a *weighted* sum of the two objectives or optimize a single objective with constraints on the allowable values of the second objective.

The *weighting* method combines the two objectives. In this case, we would have:

$$
\textit{Min} \quad w \underbrace{\sum_{i \in I} \sum_{j \in J} h_i d_{ij} Y_{ij}}_{\substack{\text{demand-weighted} \\ \text{total distance}}} + (1-w) \underbrace{\sum_{i \in I} h_i (1 - Z_i)}_{\text{uncovered demand}} \qquad \text{weighted objective function}
$$

In this weighted objective we place a weight of w on the demand-weighted total distance and a weight of $(1 - w)$ on the total uncovered demand. Note that maximizing the covered demand is the same as minimizing the uncovered demand. For values of w close to 1, the objective function essentially minimizes the demand-weighted total distance; for values close to 0, the objective minimizes the uncovered demand or, equivalently, maximizes the covered demand.

A second approach is to convert all but one of the objectives into constraints and to optimize the remaining objective. In this case, we will minimize the first objective or the demand-weighted total distance and constrain the covered demand to be at least some value, $h^c_{\min}$. In particular, we drop the second objective and add the following constraint to the problem:

$$\sum_{i \in I} h_i Z_i \geq h^c_{\min} \qquad \text{must cover at least } h^c_{\min} \text{ demands}$$

The following algorithm, or step-by-step procedure, will enable us to find all non-dominated solutions.

Step 0:   Set $h^c_{\min} = 0$.

Step 1:   Solve the following optimization problem:

$$
\begin{aligned}
Min \quad & \sum_{i \in I}\sum_{j \in J} h_i d_{ij} Y_{ij} && \text{demand-weighted total distance} \\
s.t. \quad & \sum_{j \in J} Y_{ij} = 1 && \forall i \in I && \text{each node is assigned} \\
& Y_{ij} - X_j \leq 0 && \forall i \in I, j \in J && \text{assign only to open facilities} \\
& \sum_{j \in J} X_j = p && && \text{locate } p \text{ facilities} \\
& Z_i - \sum_{j \in J} a_{ij} X_j \leq 0 && \forall i \in I && \text{linkage constraint} \\
& \sum_{i \in I} h_i Z_i \geq h^c_{\min} && && \text{must cover at least } h^c_{\min} \text{ demands} \\
& X_j \in \{0, 1\} && \forall j \in J && \text{integrality} \\
& Y_{ij} \geq 0 && \forall i \in I, j \in J && \text{non-negative variables} \\
& 0 \leq Z_i \leq 1 && \forall i \in I && \text{bounds on coverage variables}
\end{aligned}
$$

Step 2:   If a feasible solution exists to the problem of Step 1, record the new non-dominated solution with coordinates

$$\left(\frac{\sum_{i \in I} h_i Z_i}{\sum_{i \in I} h_i}, \frac{\sum_{i \in I}\sum_{j \in J} h_i d_{ij} Y_{ij}}{\sum_{i \in I} h_i} \right),$$

set $h^c_{\min} = \sum_{i \in I} h_i Z_i + 1$, and go to Step 1; otherwise, *stop*, all non-dominated solutions have been found.

Two observations are worth noting about the algorithm. First, in Step 2, we are recording the *percent of the total demand that is covered* and the *demand-weighted average distance* as the

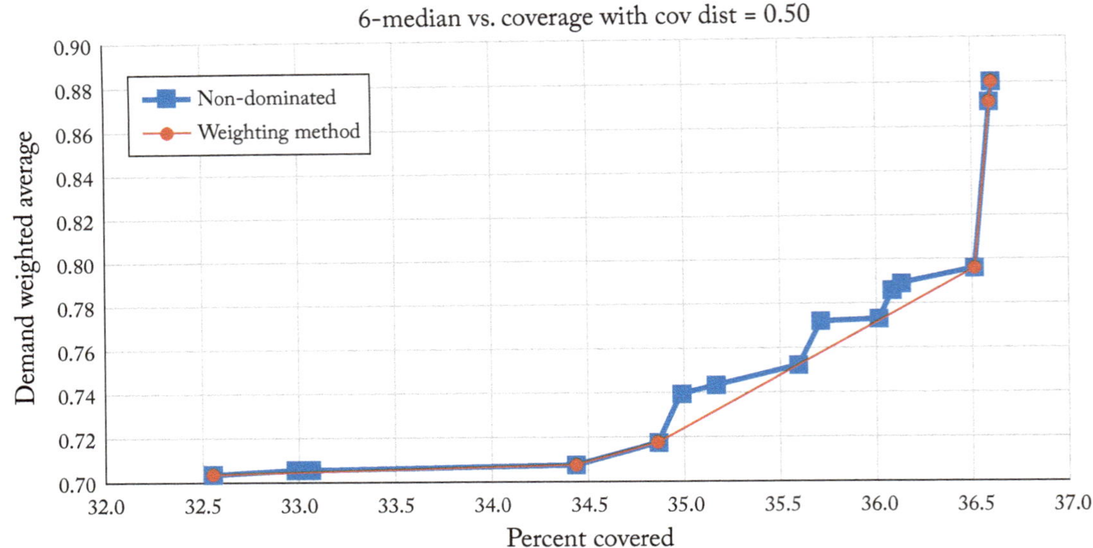

Figure 16.3: Median-covering tradeoff curve for the Ann Arbor Dataset with 6 facilities and a coverage distance of 0.5 miles.

coordinates of the non-dominated solution. Second, when we set $h^c_{\min} = \sum_{i \in I} h_i Z_i + 1$ and return to Step 1, the formerly optimal solution will no longer be feasible since the constraint that states $\sum_{i \in I} h_i Z_i \geq h^c_{\min}$ will no longer be feasible.

Figure 16.3 shows the results of using this analysis for the 33-node Census tract representation of Ann Arbor with 6 facilities and a coverage distance of 0.5 miles. The algorithm above finds the 15 solutions shown with blue squares. The lines connecting these points are there simply to help show the solutions; no solution exists along the lines except at the points represented as squares. The red line shows the five solutions we would be able to find using the *weighted* objective function above, changing the weights appropriately. Note that the weighting method does not find all of the non-dominated solutions, while the constraint method shown in the algorithm above does find all such solutions.

It is worth remembering that each solution shown in Figure 16.3 corresponds to a different set of location decisions. The solution shown in the lower left-hand corner with coordinates (32.6%, 0.703 miles) corresponds to the solution found in Figure 16.1, while the solution shown in the top right with coordinates (36.6%, 0.879 miles) corresponds to the solution shown in Figure 16.2. Thirteen other good solutions exist. In particular, Figure 16.4 shows the solution that corresponds to coordinates (34.4%, 0.708 miles). Note that this represents a 0.7% degradation in the demand-weighted average compared to the solution shown in Figure 16.1, but the percent of demand that is covered increases by 5.7%. Similarly, Figure 16.5 shows the solution that

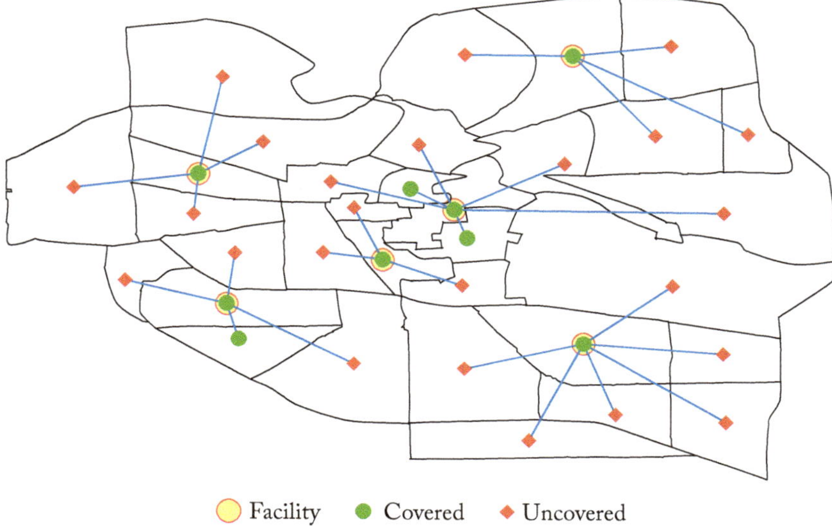

Figure 16.4: A good compromise solution.

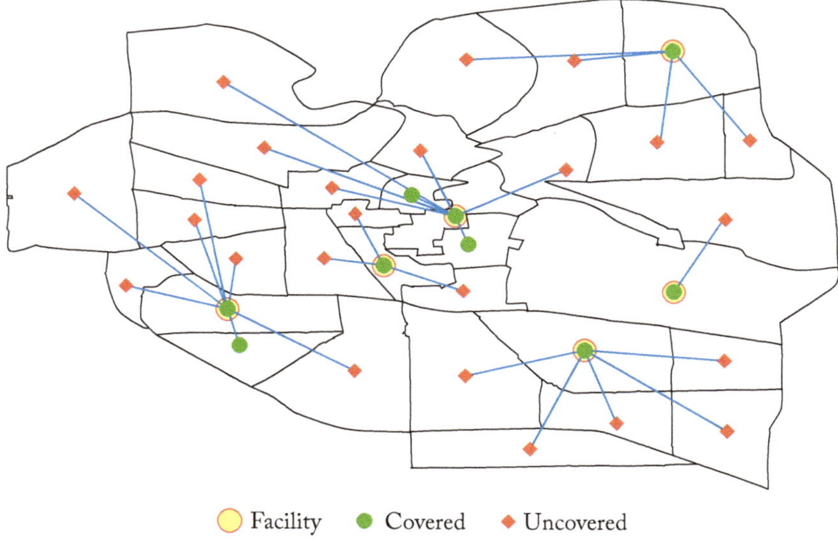

Figure 16.5: Another good compromise solution.

corresponds to coordinates (36.5%, 0.795 miles). This solution improves the demand-weighted average distance by 9.6% compared to the solution shown in Figure 16.2 while degrading the percent of the total demand that is covered by only 0.25%. Compared to the solution shown in Figure 16.4, this solution moves a facility from the west side of the city to the east side of the city.

Clearly, good compromise solutions exist in this case. This will generally be true for most multi-objective problems in location modeling and throughout operations management. The reader interested in exploring these concepts further is referred to the excellent seminal text on multi-objective modeling by Cohon [1978].

16.3   REFERENCES

J. Cohon, *Multiobjective Programming and Planning*, Academic Press, New York, 1978. DOI: 10.1007/978-3-642-45527-8_27. 118

C H A P T E R 17

Fundamentals of Queueing Theory

17.1 WHAT IS QUEUEING THEORY AND WHY DO WE CARE ABOUT IT?

Queueing theory is simply a mathematical description of the performance of waiting lines. When you call a customer service line, for example when your flight is cancelled and you need to be rebooked on another flight, you are joining a queue, or waiting line, for service. When you purchase groceries at a grocery store, you must join a queue to pay for the items you purchased. If you are like me, you inevitably join a queue that seems to be short but that proves to have long waiting times as a customer ahead of you can't find his credit card or another customer decides she really only needs two jars of tomato sauce and not three. When you go through a TSA (Transportation Security Administration) checkpoint, you must join three queues: the first to have your identification checked, the second to have your baggage x-rayed, and a third to have your body scanned using a magnetometer or another x-ray machine. There is often waiting involved in each of these three processes. When you buy your morning coffee, you often must wait in a line to order and pay for your coffee and then wait again for a barista to make your drink. That is a particularly onerous line since you clearly have not yet had your morning dose of caffeine! As parts move through an auto assembly plant, they must often wait for processing on the next machine before they eventually emerge as a gleaming new vehicle. All of these are examples of queues, which can be analyzed using queueing theory. As I write this, there are long queues of people who have ordered automobiles and other vehicles. These people are waiting for their cars to be scheduled for assembly and eventual delivery. The long queues today are caused by significant part shortages, particularly shortages of semiconductors.

In short, life is full of queues. Through our ability to analyze queues and to predict the number of people or items in a queue and the average time spent waiting for service and the average time spent in the system, we can begin to *design* queues to be more effective and efficient. Much of the remainder of this text is concerned with queueing theory. In this chapter, we introduce some basic performance metrics and some key relationships between these metrics. In Chapter 18, we show how to compute these metrics for a single-server queue with a particular inter-arrival time distribution and a particular service time distribution. In Chapter 19, we again analyze a single server queue, but allow for (just about) any general service time distribution. This will allow us to explore the impact of service time variability on the performance of queueing systems. Chapter 20 returns to the more restrictive assumption about service times, but allows for multiple servers. Chapter 21 returns to the restrictive single server problem but shows how we can use other techniques—numerical analysis and simulation—to analyze the performance of the system.

17.2   BASIC NOTATION, QUEUEING INPUTS, AND OUTPUTS

To analyze any queueing problem, we need to know at least three key inputs.

1. A characterization of the **inter-arrival** time distribution or the probability distribution of the time between successive customer arrivals. We will often assume that the inter-arrival times are *exponentially* distributed, which means that the number of arrivals in any time period follows a *Poisson* distribution. This is convenient since the exponential distribution is *memoryless*, meaning that we do not need to keep track of how long it has been since the last arrival; the distribution of time until the next arrival is independent of how long it has been since the last arrival if the inter-arrival time distribution is exponential.

2. A characterization of the **service** time distribution or the probability distribution associated with the time it takes a customer to be served once s/he begins being served. Again, we will often assume that the service time distribution is exponential.

3. The **number of servers**.

We generally summarize the three key inputs using *Kendall's queueing notation* as follows. The inter-arrival time distribution and service time distribution are summarized using letters. We use the following notation:

M	memoryless or exponentially distributed or Poisson
GI or G	general independent (used for the inter-arrival time distribution) or general (used for the service time distribution). Even though we do not explicitly state that the service times are independent, we will be making this assumption
D	deterministic
E_k	Erlang-k

The third key parameter in Kendall's notation specifies the number of servers. The general form of the notation is:

Inter-arrival time distribution/service time distribution/number of servers.

Thus, an M/M/1 queue has exponentially distributed inter-arrival times, exponentially distributed service times, and a single server. This is the simplest possible queue to analyze and is the topic of Chapters 18 and 21. An M/G/1 queue has exponentially distributed inter-arrival times, (just about) any general service time distribution, and a single server. This is the topic of Chapter 19. An M/M/s queue has exponentially distributed inter-arrival times, exponentially distributed service times, and s servers. This is the topic of Chapter 20.

In addition to these three key assumptions or input conditions, we sometimes need to deal with other inputs or characterizations of the queue. For example, some queues operate with *priority* service systems. When you board an airline, the platinum executive class passengers get to board before the rest of us. In an emergency room, someone in acute cardiac arrest will be treated before a patient who may have a sprained ankle or even a broken leg. Another input is sometimes the size of the population that is being served. A computer repair technician may have a finite number of computers that she maintains. Clearly, if a lot of those computers are broken at some point in time, the rate at which additional computers break down is going to be less than it would be if all of the computers were working. Finally, there are some systems in which a finite number of customers are allowed in the system. A municipal parking lot is an example of such a system since the lot can only accommodate a given number of vehicles at any time. The analysis of queues with these characteristics—priority service systems, a finite population, or a finite capacity—is generally beyond the scope of this text though Chapter 21 introduces a queue with a maximum capacity. Interested readers are referred to the excellent texts by Gross and Harris [1985] and White, Schmidt, and Bennett [1975] or to Chapters 3 and 9 of Daskin [2010].

We are generally interested in (at least) four key performance metrics:

L the average number of customers in the system including those waiting for service and those being served

W the average time a customer spends in the system including the time spent waiting for service and the time spent being served

W_q the average time a customer spends waiting for service to begin

L_q the average number of customers waiting for service to begin

When we talk about these performance metrics, we are generally talking about the *steady state* performance of the system. What this means is that all of the input conditions are constant and independent of time, and that we are talking about the long-run average performance of the system. While the assumption of constant inputs may not be strictly met in many situations, we can still use queueing theory to get an approximate idea of how the system performs. For example, the arrival rate of customers at a local coffee shop clearly depends on the time of day with a peak in the morning. The coffee shop is also likely to be closed for some number of hours each day (e.g., perhaps 9 pm until 5 am the following day). This means that there are no arrivals during the period of time the store is closed. Despite the violation of the assumption that the arrival rate is constant and independent of the time of day, we can still use queueing theory to get some idea of the performance of the system as we change key inputs, including the number of baristas on duty at any time of day. It is worth remembering that Box once said, "All models are wrong, but some are useful."

17.3   LITTLE'S LAW AND OTHER KEY RELATIONSHIPS

The four output measures listed above—L, W, W_q, and L_q—are intimately related. To explore the relationships between these quantities, we need to define two inputs:

λ the arrival rate of customers (customers/unit time)

$\dfrac{1}{\mu}$ the mean or average service time

With this additional notation, we can write down Little's Law which states

$$L = \lambda W$$

and

$$L_q = \lambda W_q.$$

In addition, we know that for (most) queues,

$$W = W_q + \frac{1}{\mu}.$$

(One example of a case in which the last relationship will not hold is when customers join a queue and then leave before they are served. This happens when you call a customer service desk and get tired of listening to how important your call is to them and you hang up before talking to a customer service representative. It also happens when you go to your local coffee shop for your morning dose of caffeine and are forced to leave before you get to order so that you do not risk missing the bus to your office. These cases are beyond the scope of this book.)

What this means is that if we know the arrival rate, λ, and the mean service time, $\frac{1}{\mu}$, we can compute any of the four key performance metrics once we know one of them. For example, suppose the arrival rate of customers at your local coffee shop is 60 customers per hour and the mean service time is 2 minutes of 1/30th of an hour. If, on average, there are 10 customers in the system (waiting for service and being served), then we have:

$$\lambda = 60 \text{ customers/hour}$$

$$\frac{1}{\mu} = \frac{1}{30} \text{ hour}$$

$$L = 10$$

$$W = \frac{L}{\lambda} = \frac{10}{60} = 0.1667 \text{ hours} = 10 \text{ minutes}$$

$$W_q = W - \frac{1}{\mu} = 0.1667 - 0.0333 = 0.1333 \text{ hours} = 8 \text{ minutes}$$

$$L_q = \lambda W_q = 60 \cdot 0.1333 = 8 \text{ people}$$

Note that the first three quantities are given and the last three are computed using those quantities. In subsequent chapters we will discuss how to compute at least one of these quantities so that we can then use the relationships shown above to compute the other four key performance metrics.

Finally, it is worth noting that there are often other performance metrics of interest. For example, we may be interested in the probability that a customer will have to wait for service. We may want to know the probability that a customer will have to wait more than some number of minutes. We may want to know the probability that there will be more than some number of customers in the system. Computing some of these metrics will be relatively easy, while computing others will be beyond the scope of this text.

17.4 REFERENCES

M. S. Daskin, *Service Science*, John Wiley & Sons, New York, 2010. DOI: 10.1002/9780470877876. 121

D. Gross and C. Harris, *Fundamentals of Queueing Theory*, 2nd ed., John Wiley & Sons, New York, 1985. DOI: 10.1002/9781118625651. 121

J. A. White, J. W. Schmidt, and G. K. Bennett, *Analysis of Queueing Systems*, Academic Press, New York, 1975. DOI: 10.1016/B978-0-12-746950-8.X5001-9. 121

C H A P T E R   18

A Single-Server Queue

18.1   A VERY SIMPLE QUEUE

The simplest possible queue we can analyze is the M/M/1 queue. Recall that this is a queue with exponentially distributed inter-arrival times (or Poisson arrivals), exponentially distributed service times, and a single server. While the characteristics of many real-world queues may differ from the assumptions made in this queueing model, some of the fundamental principles associated with queueing theory will carry over to more complex systems as well. By studying this queue first, we will be able to understand how to analyze similar queueing problems, including the more complex problem of Chapter 20 with multiple servers. We will derive some key insights into the behavior of just about every queueing system we might encounter.

With the assumptions of exponentially distributed inter-arrival times and exponentially distributed service times, the system becomes memoryless or Markovian. This means that we do not need to know how long it has been since the last arrival or how long the customer who is currently being served has been with the server. The conditional distributions of (1) the time until the next arrival and (2) the time until the current customer finishes her service are both exponential with the same parameter as the unconditional inter-arrival time distribution and the service time distribution, respectively. We will let λ be the arrival rate and μ be the service rate. This means that the mean service time is $\frac{1}{\mu}$ and the mean time between arrivals is $\frac{1}{\lambda}$. It should be intuitively clear that we will require $\lambda < \mu$, since otherwise people will be arriving at a rate (equal to or) greater than the rate at which they can be served. This will mean that the number in the system grows with time, on average. Note that we need to exclude the case in which the arrival rate equals the service rate since the randomness in the system would lead to instability here as well.

We can represent this simple queue with the state transition diagram shown in Figure 18.1. The state of the system represents the number of customers in the system including those being served and those waiting for service. The circles in the figure represent the state of the system. State 0 means that there is no one in the system. State 1 means that there is one person in the system, and the server is clearly serving that person. State 2 means that there are two people in the system, one is being served and the other is waiting for service. In general, in state n, there is 1 person being served and $n - 1$ people waiting for service, provided $n \geq 1$.

The arrows in Figure 18.1 represent the possible transitions that can be made in the system. For example, the system can transition from state 0 to state 1 if there is an arrival to the system. This occurs at a rate of λ. Similarly, the system can transition from any state $n \geq 1$ to state $n + 1$

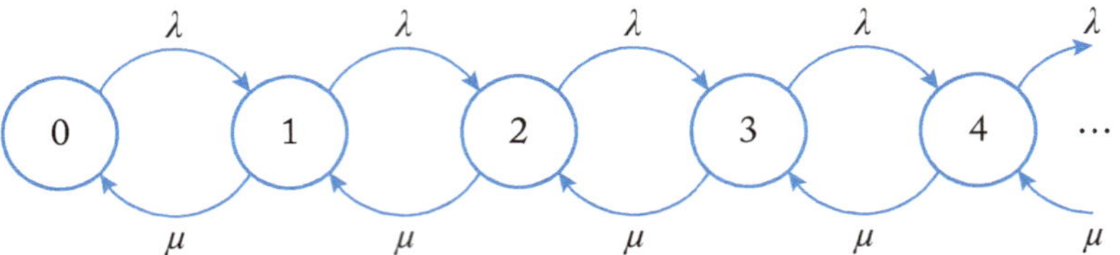

Figure 18.1: State transition diagram for an M/M/1 queue.

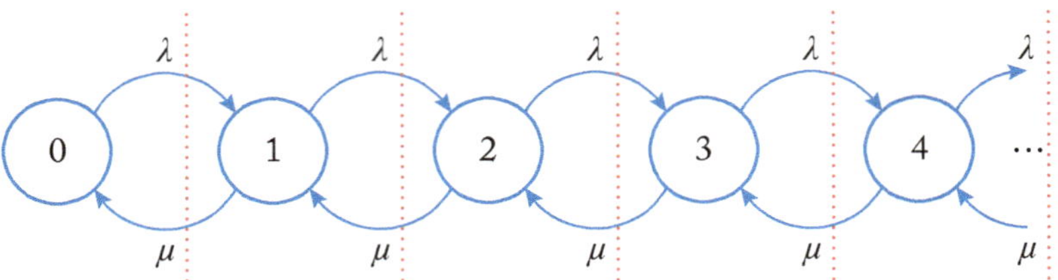

Figure 18.2: State transition diagram for steady state analysis.

at rate λ. Service to customers is completed at rate μ. Therefore the system can transition from state n to state $n-1$ at rate μ, for any state $n \geq 1$. Finally, it is worth noting that while we have shown only the first five states (states 0 through (4) in the figure, the state space extends to infinity; there are an infinite number of states in the system.

18.2 STEADY STATE ANALYSIS OF THE M/M/1 QUEUE

Let us now turn our attention to Figure 18.2. This is identical to Figure 18.1 except that we have now drawn vertical dashed red lines between each successive pair of states. These lines will enable us to write down the steady state balance equations for this queue. In particular, in steady state, we require that the rate of probability flux from left to right across one of these dashed lines must equal the rate of probability flux from right to left across the same line. By *probability flux* we mean the probability of being in the state corresponding to the tail of the arrow from that state times the rate at which the corresponding transition occurs. Thus, the rate of probability flux from left to right across the dashed line between states 0 and 1 is given by λP_0, where P_n is the probability that the system is in state n. (Note that this notation differs slightly from that used in Chapter 4. In Chapter 4, uppercase P was used to denote the cumulative probability. Here, and in subsequent chapters on queueing theory, P_n denotes the probability that the system is in state n. It is not a cumulative probability.) Similarly, the rate of probability flux across this line from right to left is given by μP_1. In steady state, we require $\lambda P_0 = \mu P_1$ or equivalently, $P_1 = \frac{\lambda}{\mu} P_0$.

For the line between states 1 and 2, we would require $\lambda P_1 = \mu P_2$ or $P_2 = \frac{\lambda}{\mu} P_1$. But since $P_1 = \frac{\lambda}{\mu} P_0$, we can write $P_2 = \frac{\lambda}{\mu} P_1 = \left(\frac{\lambda}{\mu}\right)^2 P_0$. Similarly, we can show that we will have $P_n = \frac{\lambda}{\mu} P_{n-1} = (\frac{\lambda}{\mu})^n P_0$. Thus, we can write the probability of being in any state in terms of the probability of being in state 0.

Let us now let $\rho = \frac{\lambda}{\mu} \cdot \rho$ is the utilization rate of the server in this case. We can now write $P_n = \rho^n P_0$. We also know that the (infinite) sum of all the state probabilities must equal 1, or $\sum_{n=0}^{\infty} P_n = \sum_{n=0}^{\infty} \rho^n P_0 = P_0 \sum_{n=0}^{\infty} \rho^n = 1$. But, for $\rho < 1$, the infinite sum simply equals $\frac{1}{1-\rho}$, which, in turn, means that $P_0 = 1 - \rho$ and $P_n = (1 - \rho)\rho^n$ for $n = 0, 1, \ldots$.

But this is nothing more than the geometric distribution of Section 4.5 if we let $q = 1 - \rho$, where q was the probability of success in the Bernoulli trial underlying the geometric distribution.

We can now compute the key performance metrics for this queue. We obtain:

$$L = \sum_{n=0}^{\infty} n(1 - \rho)\rho^n = \frac{\rho}{1 - \rho}$$

$$W = \frac{L}{\lambda} = \frac{1}{\mu(1 - \rho)}$$

$$W_q = W - \frac{1}{\mu} = \frac{1}{\mu(1 - \rho)} - \frac{1 - \rho}{\mu(1 - \rho)} = \frac{\rho}{\mu(1 - \rho)}$$

$$L_q = \lambda W_q = \frac{\lambda \cdot \rho}{\mu(1 - \rho)} = \frac{\rho^2}{(1 - \rho)}.$$

It is worth noting that each of these terms has $1 - \rho$ in the denominator. This means that as the arrival rate approaches the service rate (as $\rho \to 1$), the performance of the system degrades considerably, no matter which performance metric we are considering. We can also show that

$$Var(N) = \frac{\rho}{(1 - \rho)^2}$$

$$SD(N) = \frac{\sqrt{\rho}}{1 - \rho},$$

where N is a random variable denoting the number of customers in the system. This means that not only does the performance degrade as the utilization increases, but the *variability* of the performance, as measured by either the variance of the number in the system, $Var(N)$, or by the standard deviation of the number in the system, $SD(N)$, also increases as the utilization increases.

Figure 18.3 plots the performance metrics for the M/M/1 queue as a function of the utilization ratio when the mean service time is equal to 1, or $\frac{1}{\mu} = 1$. No matter which metric we are plotting, the performance degrades as the utilization ratio gets larger. For utilization ratios great than 0.9, the performance of the queue degrades significantly. Thus, for utilization ratios

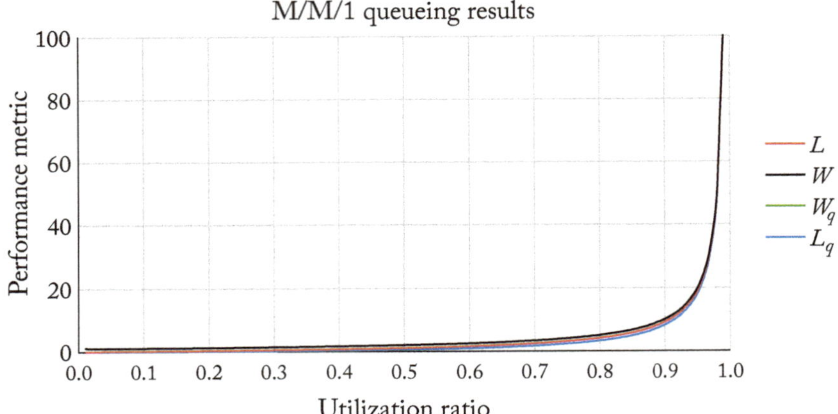

Figure 18.3: Performance of the M/M/1 queue as a function of the utilization ratio.

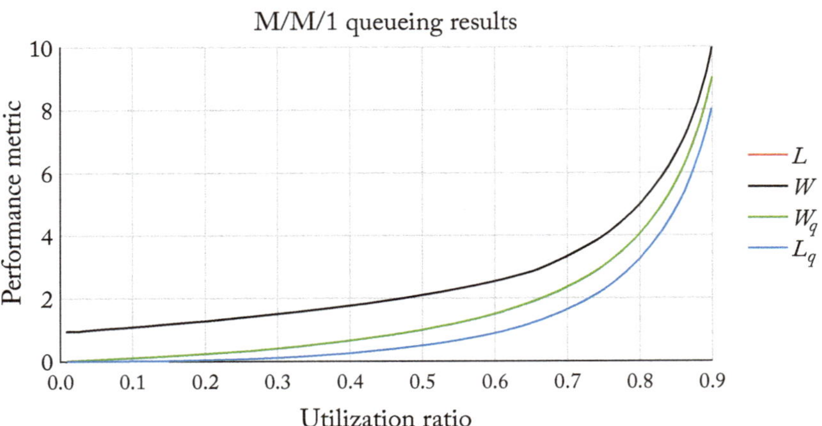

Figure 18.4: M/M/1 performance with the utilization ratio less than or equal to 0.9.

less than 0.5, the mean time in the system is at most twice the service time. If the utilization ratio is 0.75, the average time in the system doubles to four times the mean service time. When the utilization ratio is 0.9, the mean time in the system is 10 times the mean service time; with a utilization ratio of 0.95, the mean time in the system is twice the value when the utilization ratio is 0.9.

Figure 18.4 plots the performance of the system for utilization ratios less than 0.9. Here it is easier to see the different performance metrics, though when $\mu = 1$, the average number in the system, L, is always equal to the average time waiting for service, W_q, making it impossible to distinguish these two curves in the figure.

Finally, let us look at a numerical example. Suppose arrivals at a small local coffee shop are Poisson with a rate of $\lambda = 20$ per hour and that the mean service time is $\frac{1}{\mu} = \frac{1}{30}$ hours or 2 minutes. In this case, we have $\mu = 30$ and $\rho = 2/3$. We then have $L = 2$ people, $W = 6$ minutes, $W_q = 4$ minutes, and $L_q = 4/3$ people.

CHAPTER 19

A More General Single-Server Queue and The Impact of Variability on Queueing Performance

19.1   A QUEUE WITH GENERAL SERVICE TIMES

The assumption of Poisson arrivals or exponential inter-arrival times is often (largely) satisfied, at least for many service systems. It is less often satisfied for manufacturing contexts in which parts often arrive at a processing center according to a regular schedule. For example, a new part may arrive at a processing center every 5 seconds. This would hardly be an exponential distribution of inter-arrival times; in fact, this would be a deterministic inter-arrival time distribution.

In this chapter, however, we will focus on the service time distribution. We will relax the even more restrictive assumption that the service times are exponentially distributed. While this assumption makes analyzing queueing systems and deriving closed form results relatively easy, it is often not supported by data in the real world. Thus, it is important to relax this assumption.

Fortunately, it is relatively easy to do so. We will consider a single-server queue with Poisson arrivals or exponentially distributed inter-arrival times, and any general service time distribution with a finite mean, $\frac{1}{\mu}$, and a finite variance, σ^2. (While most of the distributions we think of satisfy this condition, there are some distributions, like the Cauchy distribution, which do not have a finite variance. For our purposes, we do not need to worry about such distributions.)

Under these relatively mild conditions, we can show that the average number of customers in the system is given by:

$$L = \rho + \frac{\rho^2 + \lambda^2\sigma^2}{2(1-\rho)},$$

where $\rho = \frac{\lambda}{\mu}$ and we again require $\rho < 1$ for steady state performance of the system. This formula is known as the *Pollaczek–Khinchine formula*. The reader is referred to Clarke and Disney [1970] and Gross and Harris [1985] for a derivation of this formula.

This is clearly a generalization of the formula that we derived in Chapter 18 for the average number of people in an M/M/1 queue. One of the first things we should do whenever we encounter a new formula that is a generalization of an earlier formula is to be sure that the new formula gives us the earlier results when we substitute the conditions that applied for the earlier, less general, formula. In this case, we simply need to substitute the variance of an exponential service time distribution with mean $\frac{1}{\mu}$. This variance is $\sigma^2 = \frac{1}{\mu^2}$. When we substitute this value of the variance into the Pollaczek–Khinchine formula, we obtain:

$$L = \rho + \frac{\rho^2 + \lambda^2\sigma^2}{2(1-\rho)} = \rho + \frac{\rho^2 + \lambda^2/\mu^2}{2(1-\rho)} = \rho + \frac{2\rho^2}{2(1-\rho)} = \frac{2\rho - 2\rho^2 + 2\rho^2}{2(1-\rho)} = \frac{\rho}{1-\rho}.$$

In other words, the Pollaczek–Khinchine formula reverts to the well-known result for the M/M/1 queue that we derived in Chapter 18, as expected. Clearly, this is good news!

We can now use the relationships that we outlined in Chapter 17 to derive the remaining three fundamental performance metrics for the M/G/1 queue. In particular, we have:

$$W = \frac{L}{\lambda} = \frac{1}{\mu} + \frac{\rho^2 + \lambda^2\sigma^2}{2\lambda(1-\rho)}$$

$$W_q = W - \frac{1}{\mu} = \frac{\rho^2 + \lambda^2\sigma^2}{2\lambda(1-\rho)}$$

$$L_q = \lambda W_q = \frac{\rho^2 + \lambda^2\sigma^2}{2(1-\rho)}.$$

19.2   THE IMPACT OF SERVICE TIME VARIABILITY ON PERFORMANCE

One of the nice features of the Pollaczek–Khinchine formula is that it explicitly tells us how the variance of the service time distribution, σ^2, impacts the performance of the system. We will focus on the average time spent waiting for service, W_q. In particular, we will compare the waiting time for the M/M/1 queue, which we will denote here as $W_q^{M/M/1}$, to the waiting time for the M/D/1 queue which has deterministic service times. We will denote this waiting time as $W_q^{M/D/1}$. For the M/D/1 queue, we have $\sigma^2 = 0$, while for the M/M/1 queue $\sigma^2 = \frac{1}{\mu^2}$.

Using these variance formulae, we can derive

$$W_q^{M/M/1} = \frac{\rho}{\mu(1-\rho)}$$

$$W_q^{M/D/1} = \frac{\rho}{2\mu(1-\rho)} = \frac{1}{2}W_q^{M/M/1}.$$

In particular, the waiting time in the queue with deterministic service times is exactly one half of the average waiting time in a queue with exponential service times. **Clearly, variability hurts us.** The same relationships occur when we look at the average number of people waiting for service in the two queues. We obtain:

$$L_q^{M/M/1} = \lambda W_q^{M/M/1} = \frac{\rho^2}{(1-\rho)}$$

$$L_q^{M/D/1} = \lambda W_q^{M/D/1} = \frac{\rho^2}{2(1-\rho)} = \frac{1}{2} L_q^{M/M/1}.$$

Again, the average number of people waiting for service in an M/D/1 queue will be exactly half the average number waiting in an M/M/1 queue with identical arrival rates and mean service times.

As a final note, let us consider what happens if the service times follow an *Erlang-k* distribution with mean $\frac{1}{\mu}$ and variance $\sigma^2 = \frac{1}{k\mu^2}$. In this case we obtain:

$$L = \rho + \frac{\rho^2 + \lambda^2\sigma^2}{2(1-\rho)} = \rho + \frac{\rho^2 + \lambda^2/k\mu^2}{2(1-\rho)} = \rho + \frac{\left(1 + \frac{1}{k}\right)\rho^2}{2(1-\rho)}$$

$$W = \frac{L}{\lambda} = \frac{1}{\mu} + \frac{\left(1 + \frac{1}{k}\right)\rho^2}{2\lambda(1-\rho)}$$

$$W_q = W - \frac{1}{\mu} = \frac{\left(1 + \frac{1}{k}\right)\rho^2}{2\lambda(1-\rho)}$$

$$L_q = \frac{\left(1 + \frac{1}{k}\right)\rho^2}{2(1-\rho)}.$$

Clearly, as the variance of the Erlang-k distribution decreases (or as k increases), the performance of the queue improves, as expected.

Finally, let us consider a simple numerical example. We let the mean service time be equal to 1 and the arrival rate equal 0.8. The utilization ratio is therefore 0.8. Table 19.1 summarizes the results for the four key performance metrics and five different service time distributions. As expected the average time in the queue and the average number in the queue when the service time distribution is deterministic are exactly half of the corresponding values when the service times are exponentially distributed. **Reducing the variability (variance) of service times will lead to performance improvements in queueing systems.**

Table 19.1: Key performance metrics for an M/G/1 queue with $\rho = 0.8$ and different service time distributions

Performance Metric	Service Time Distribution				
	Exponential	Erlang-2	Erlang-4	Erlang-9	Deterministic
L	4.00	3.20	2.80	2.58	2.40
W	5.00	4.00	3.50	3.22	3.00
Wq	4.00	3.00	2.50	2.22	2.00
Lq	3.20	2.40	2.00	1.78	1.60

19.3 REFERENCES

A. B. Clarke and R. L. Disney, *Probability and Random Processes for Engineers and Scientists*, John Wiley & Sons, New York, 1970. 131

D. Gross and C. Harris, *Fundamentals of Queueing Theory*, 2nd ed., John Wiley & Sons, New York, 1985. DOI: 10.1002/9781118625651. 131

C H A P T E R   20

A Simple Multi-Server Queue

20.1   WHAT IF WE HAVE MULTIPLE SERVERS?

Some service systems operate as if they had a single server. My grandfather was a butcher and owned a small store in New Jersey. I doubt that he had any employees and so his store operated as a single-server queue. The models of Chapters 18 and 19 will provide some insights into the operation of such a store.

Most systems, however, have more than one server. The days of the local butcher are largely gone, at least in most of the United States. Instead, many of us buy pre-packaged meat and poultry and fish from a large supermarket. Fresh meats and fish are also sold in supermarkets, but there are often multiple individuals staffing the meat department or the fish department. Similarly, when you call an airline for rebooking after a flight was cancelled, you are calling a large call center with many customer service agents [Gans, Koole, and Mandelbaum, 2003]. When you pull up to a gas station, there are often half a dozen or more pumps, each of which is a server.

In this chapter, we will explore the impact of having multiple servers. Before diving further into the analysis of multi-server queues, however, we need to state clearly what form such a queue takes. Figure 20.1 shows a queue with three servers, shown as yellow rounded squares. There are eight customers in the system as shown by the blue circles in the figure: five are waiting for service **in a single line** and three are being served, one by each of the three servers. When a customer finishes his/her service, the server will then attend to one of the waiting customers, generally the first customer in the line.

This form of a queue is typical of the way in which customers in a post office are served. There is one long line for service and the patron at the front of the line is the next served by one of the postal agents. Similarly, there is one line for service at a TSA identification checkpoint. This single line for service model must be distinguished from having a line for each server. This is typical of many grocery stores where you must choose which cashier's line to join. Such a system is shown in Figure 20.2. (I recently noticed that the Trader Joe's grocery store at the corner of Atlantic Avenue and Court Street in Brooklyn, NY is set up with a system like that shown in Figure 20.1. Apparently many other grocery stores in New York are configured similarly.) At a TSA checkpoint, you often have to select a line for your carryon baggage to be x-rayed and checked and for you to go through a magnetometer or x-ray machine. In other words, after the ID check portion of a TSA screening which operates in the form shown in Figure 20.1, the system reverts to a set of parallel single-server queues as shown in Figure 20.2.

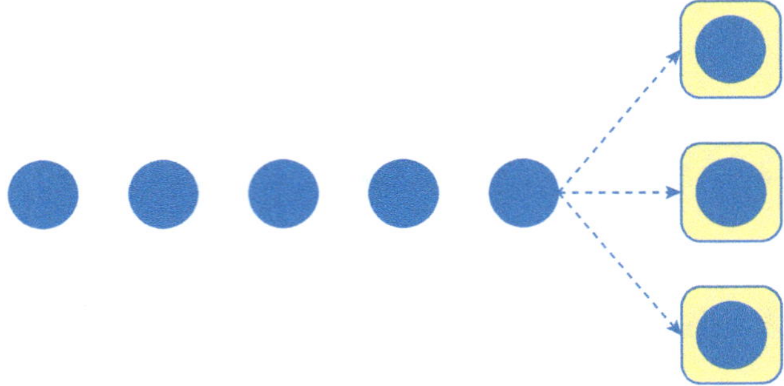

Figure 20.1: A queue with three servers.

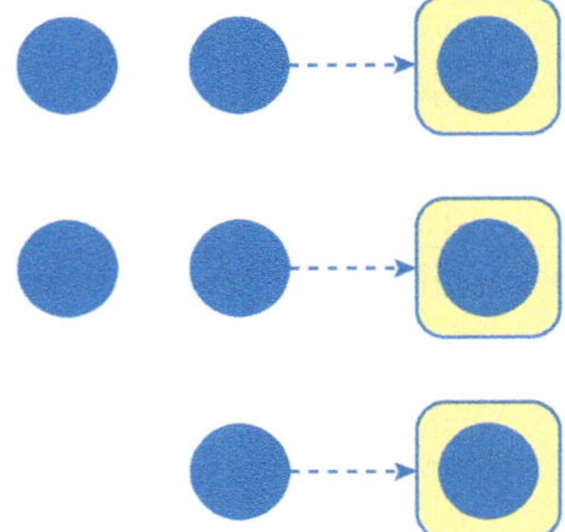

Figure 20.2: Three parallel single-server queues.

Two key questions arise. First, how do we analyze a multi-server queue of the form shown in Figure 20.1, and second, which configuration is more effective or efficient?

20.2   THE M/M/S QUEUE

In this section, we will analyze a queue of the form shown in Figure 20.1. We will go back to the assumptions of Chapter 18. In particular, we will assume that the inter-arrival times of customers are exponentially distributed, which means that the number of arrivals in a fixed time period follows a Poisson distribution. We will also assume that the service time distribution is exponentially distributed. These assumptions ensure that the system is memoryless. The rate of customer arrivals is given by λ and the service rate is given by μ. This means that the mean service time is $\frac{1}{\mu}$. There are s servers in the system.

As before, we can describe the system using a state transition diagram. Figure 20.3 shows the state transition diagram for the M/M/s queue. The upward transitions from state n to state $n + 1$ are identical to those of the M/M/1 queue. The differences occur in the downward transitions. In particular, in going from state n to state $n - 1$, the downward transition rate is equal

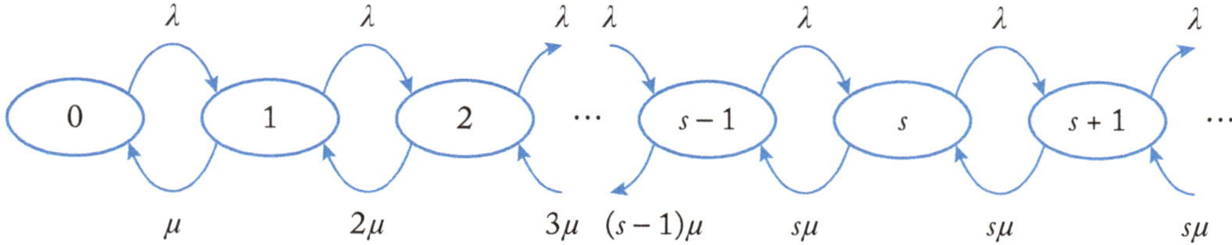

Figure 20.3: State transition diagram for an M/M/s queue.

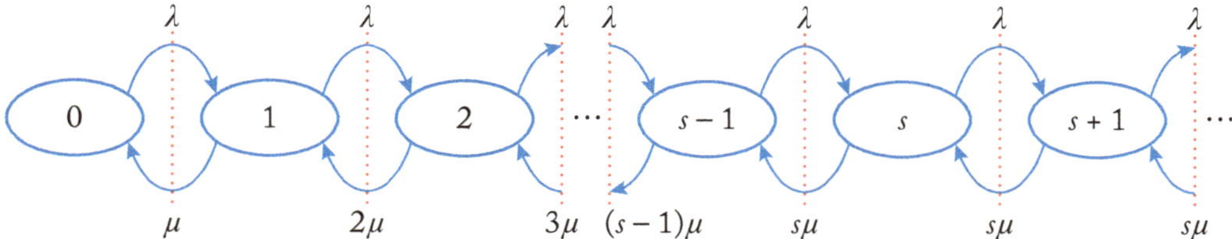

Figure 20.4: State transition diagram for an M/M/s queue for steady state analysis.

to $n\mu$, if $n \leq s$. When there are s or more customers in the system, all servers are busy and the rate at which customers leave the system is equal to $s\mu$.

As in Chapter 18, in steady state, the rate of probability flux to the right across a red dashed line in Figure 20.4 must equal the rate of probability flux to the left across the line. As before, this will lead to an infinite set of equations in which we can write the probability of being in state n, P_n, in terms of the probability of being in state 0, P_0. This can be coupled with the normalizing equation, $\sum_{n=0}^{\infty} P_n = 1$, to solve for P_0.

The result of this analysis is the following equation for P_0, the probability that the system is in the empty and idle state:

$$
\begin{aligned}
P_0 &= \left[\sum_{j=0}^{s-1} \frac{(\lambda/\mu)^j}{j!} + \frac{(\lambda/\mu)^s}{s!} \sum_{j=0}^{\infty} \left(\frac{\lambda}{s\mu} \right)^j \right]^{-1} \\
&= \left[\sum_{j=0}^{s-1} \frac{(\lambda/\mu)^j}{j!} + \frac{(\lambda/\mu)^s}{s!} \frac{s\mu}{s\mu - \lambda} \right]^{-1} \\
&= \left[e^{\lambda/\mu} \sum_{j=0}^{s-1} \frac{(\lambda/\mu)^j e^{-\lambda/\mu}}{j!} + \frac{(\lambda/\mu)^s}{s!} \frac{s\mu}{s\mu - \lambda} \right]^{-1}.
\end{aligned}
$$

Yes, this looks messy; much messier than the equation we found in Chapter 18 for the M/M/1 queue, which was simply $P_0 = 1 - \lambda/\mu$. But before we get too worried about this equation, let us break it down. In going from the second to the third line, we have multiplied the first term by $e^{\lambda/\mu}$ outside the summation and by $e^{-\lambda/\mu}$ inside the summation. These terms are shown in red. Why would we want to do this? It just makes the equation look even worse, unless you really like red equations. The reason is that the summation term is now simply the cumulative Poisson distribution with parameter λ/μ. In particular, it gives the probability that there are $s-1$ or fewer Poisson events when the Poisson rate is λ/μ. This is readily tabulated and a simple formula for this exists in Excel. This makes evaluation of P_0 relatively straightforward since we no longer need to evaluate a finite sum explicitly.

Once we know P_0, we can compute the average number of customers waiting for service. This turns out to be:

$$L_q = \frac{\lambda\mu}{(s-1)!}\frac{(\lambda/\mu)^s}{(s\mu-\lambda)^2}P_0.$$

Once we know L_q, we can compute the other three key performance metrics as shown below:

$$W_q = \frac{L_q}{\lambda}$$

$$W = W_q + \frac{1}{\mu}$$

$$L = \lambda W.$$

It is worth noting that the steady state condition for the M/M/s queue is that $\rho = \frac{\lambda}{s\mu} < 1$. In other words, we require the arrival rate, λ to be strictly less than the maximum rate at which the system can process customers, or $s\mu$. Finally, for the M/M/s queue, we can show that the probability of waiting is given by:

$$P(wait) = \sum_{j=s}^{\infty} P_j = \frac{(\lambda/\mu)^s}{s!}\left(\frac{s\mu}{s\mu-\lambda}\right)P_0$$

and the distribution of waiting before service is given by:

$$f_{W_q}(w_q) = \begin{cases} 1 - P(wait) & w_q = 0 \\ P(wait)(s\mu-\lambda)e^{-(s\mu-\lambda)w_q} & w_q > 0. \end{cases}$$

Note that this is both a discrete distribution (when the waiting time is equal to 0) and a continuous distribution (when the waiting time is strictly positive). The cumulative distribution of waiting time before service is given by:

$$F_{W_q}(w_q) = \begin{cases} 1 - P(wait) & w_q = 0 \\ P(wait)\left[1 - e^{-(s\mu-\lambda)w_q}\right] & w_q > 0. \end{cases}$$

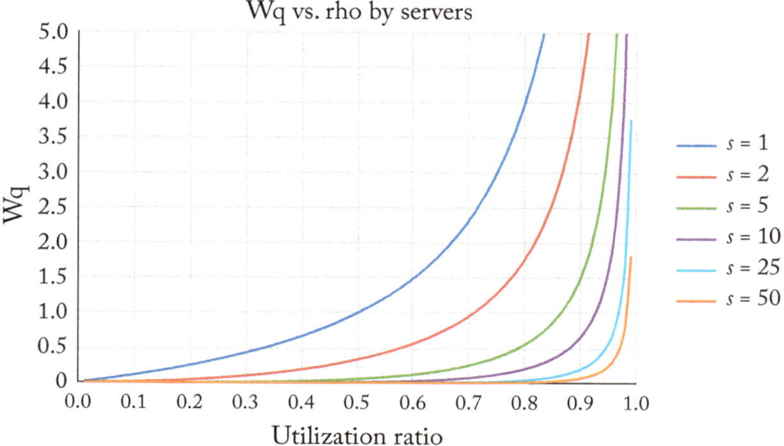

Figure 20.5: Waiting time vs. utilization ratio for six different M/M/s queues.

20.3 WHICH CONFIGURATION IS BETTER AND WHY BOTHER WITH THESE MESSY EQUATIONS ANYWAY?

Clearly, the equations we derived in Section 20.2 are significantly messier than those we derived in Chapter 18 for the M/M/1 queue. In this section, we will address two questions. First, is the configuration of Figure 20.1 or Figure 20.2 better in terms of customer service? Second, given that these equations are very messy, why bother? In particular, why not just approximate the performance of an M/M/s queue by the performance of an M/M/1 queue with either (1) the arrival rate reduced by a factor of s, since on average each server in the M/M/s queue will see and serve $1/s^{\text{th}}$ of the customers, or (2) the service rate of a single server increased by a factor of s, since when all servers are busy, people depart from the M/M/s queue at a rate of $s\mu$. In particular, we will show why either of these "approximations" of an M/M/s queue is a bad idea.

First, we compare the performance of six M/M/s queues with 1, 2, 5, 10, 25, and 50 servers. To make the comparisons fair, the system with s servers has an arrival rate that is s times that of the system with just one server. Thus, the system with 50 servers has an arrival rate that is 50 times that of the single server system for identical values of the utilization ratio. In all cases, the mean service time is 1, meaning $\mu = 1$. Figure 20.5 shows the result. The vertical axis is the waiting time in the queue, with one unit equal to one mean service time. The single server system, shown at the top of the figure exhibits the worst performance. The system with 50 servers, shown in orange, shows the best performance. Another way to view this graph is as follows. When the utilization ratio of the M/M/1 queue is 0.5, the average waiting time in the queue is one service time. For the M/M/2 queue, the utilization ratio can be as large as 0.7 before the waiting time

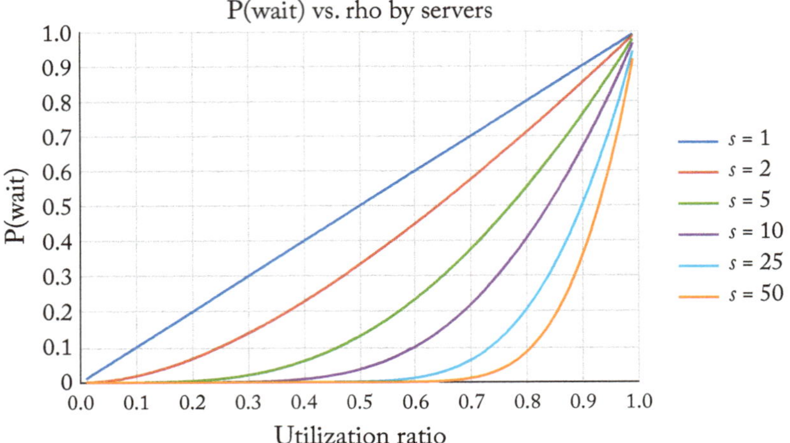

Figure 20.6: Probability of waiting vs. utilization ratio for six different M/M/s queues.

in the queue exceeds one service time. For the M/M/50 queue, the utilization ratio can be as large as 0.98 before the waiting time exceeds one service time. **Clearly, aggregating servers into one large system is advantageous from the customer's perspective.** Figure 20.6 plots the probability of waiting vs. the utilization ratio for the six different queues and Figure 20.7 plots the probability of waiting more than one service time for the six different queues. Again, we conclude that aggregating servers is beneficial from the customer's perspective. The reason for this is that the smaller systems operate independently. It is possible that one server will have a queue of customers waiting for service while another will be idle. We often see this in grocery stores.

Now let us turn to how we model such systems. Clearly, the equations for the M/M/s queue are more complex than are those of the M/M/1 queue. Why not approximate the performance of the M/M/s queue by either reducing the arrival rate or increasing the service rate so that the M/M/1 queue "looks like" the M/M/s queue? Figure 20.8 shows the results. The mean service time for the M/M/25 queue is 1 unit. The true performance is shown in blue for an M/M/25 queue. If we reduce the arrival rate by a factor of 25 and then use the formulae for an M/M/1 queue, the result is shown in red. Clearly, this overestimates the average time in the system. If we speed up the single server by a factor of 25 and then use the M/M/1 equations, we get the green line. Clearly, this result is a bit nonsensical since the resulting model predicts that the average time in the system will not exceed 1 unit until $\rho \geq 0.96$. This makes no sense since the mean service time, of which the total time is a part, is 1. Thus, this approximation consistently underestimates the performance of the M/M/25 system.

In short, you should always use the multi-server formulae when dealing with a system with multiple servers instead of trying to approximate the performance by altering the input pa-

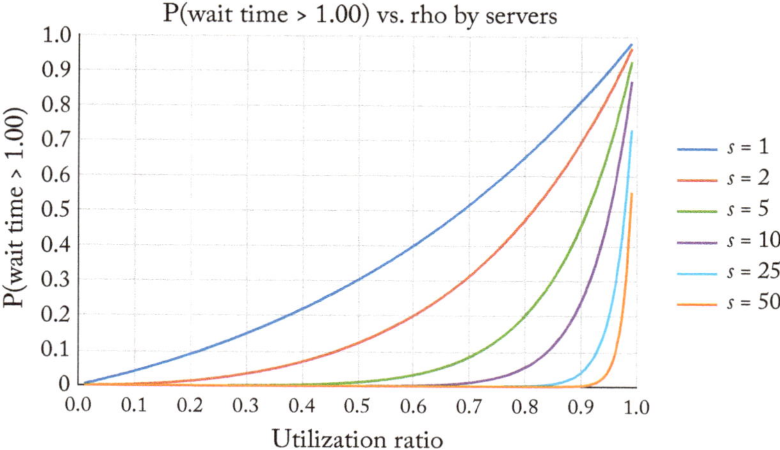

Figure 20.7: Probability of waiting more than one service time for six different M/M/s queues.

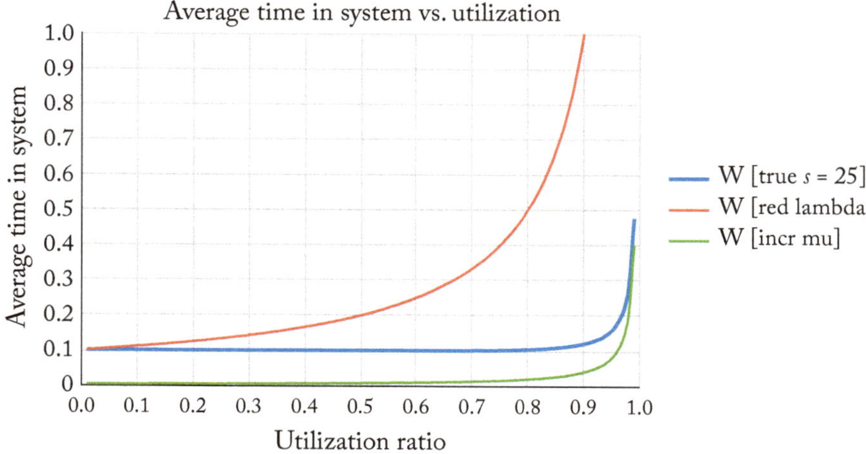

Figure 20.8: Comparison of true average time in the system and two approximations vs. utilization ratio.

rameters and employing the simpler M/M/1 equations. While the M/M/s equations are more complex, the online supplement to the text includes a spreadsheet that computes the key performance metrics for an M/M/s queue given the arrival rate, service rate, and number of servers.

Chapter 21 introduces two other methods of analyzing queues which allow us to relax some of the assumptions made in the analytic modeling of this chapter and the preceding two chapters.

20.4   REFERENCES

N. Gans, G. Koole, and A. Mandelbaum, Telephone call centers: Tutorial, review, and research prospects, *Manufacturing and Service Operations Management*, 5(2):79–141, 2003. DOI: 10.1287/msom.5.2.79.16071. 135

C H A P T E R   21

Solving Queueing Equations Numerically and Simulating the Performance of a Queue

21.1   GOING BEYOND CLOSED FORM MODELING

Chapters 18, 19, and 20 presented closed form equations for estimating the performance of simple queueing systems. Not all systems, however, have Poisson arrivals or exponentially distributed inter-arrival times, as was assumed in each of these chapters. Some systems may have scheduled customer arrivals with much less variability about the arrival times than is exhibited by Poisson arrivals. Some systems have more general service time distributions than was assumed in Chapters 18 and 20 in which we needed to assume that service times were exponentially distributed. Some systems have priorities for different customer classes (e.g., emergency patients and non-emergent individuals). In many cases, the number of servers on duty changes over the course of the day, and in some cases, in response to the state of the system. When many people are waiting to check out of a grocery store, the manager may open a new line to expedite the process. Some systems allow only a finite number of customers in the system. Parking structures are an example of this sort of queueing system. In some cases, customers may join a queue, but renege on the queue (or quit the queue) before they are served. When you hang up during a call to a call center because you are tired of hearing how important your call is to them, you are reneging on the queue. When you get in line at the campus coffee shop for a latte between classes and then realize you have to run to your next class before you are able to order, you are again reneging on the system. Finally, in some systems, there is a finite population of customers, making the arrival rate state-dependent.

In short, practical queueing systems are significantly more complex than are the simple systems analyzed in the three preceding chapters. In this chapter, we introduce two new modeling approaches that allow us to analyze more complex systems. The first is a numerical approach to solving the steady state balance equations and the second is simulation.

21.2 SOLVING THE STEADY STATE BALANCE EQUATIONS NUMERICALLY

The basic approach that we employed in deriving the closed form equations of Chapters 18 and 20 is the following.

> **Step 1**: Write down the steady state balance equations for state n in terms of the probability of being in state $n - 1$.

> **Step 2**: Solve for the state probabilities in terms of P_0, the probability that the system is in the empty and idle state.

> **Step 3**: Use the normalizing equation $\sum_{\text{all } n} P_n = 1$, to solve for P_0, and, by implication all of the other state probabilities.

> **Step 4**: Use the state probabilities to solve for the performance of the system. For example, we can obtain, $L = \sum_{\text{all } n} n P_n$.

The approach we outline in this section does exactly this, but does so numerically. But wait, you say. The systems for Chapters 18 and 20 had an infinite number of states. How can we solve an infinite number of equations numerically? In short, we will not. We will solve a finite system of equations, recognizing that the probability of large numbers of customers in a system is generally very small. For example, for an M/M/1 queue with $\rho = 0.995$, the probability of having 3,000 or more customers in the system is about 0.0000003. We do not need to worry about states beyond state 3,000 in this case for all practical purposes.

To motivate this analysis approach, consider a small urban parking lot that has 10 spaces. Whenever those spaces (or servers) are all full, customers are rejected from the system and cannot join a waiting line for service; the lot simply hangs out a sign saying "Sorry, we are full." As before, we will assume that arrivals occur according to a Poisson process (or with exponentially distributed inter-arrival times) and exponentially distributed service times. We could draw the state transition diagram similar to the ones drawn in Chapters 18 and 20. This one would have 11 states, corresponding to states 0, 1, …, 10. We could use the algorithm above to solve for the performance of the system. However, the equations would get to be messy since we have finite sums instead of infinite sums. The reader interested in this analytic approach should see Chapter 3 of Daskin [2010].

Let us consider a specific example. Customers wanting to park arrive according to a Poisson process at the rate of two per hour. The mean time a car spends in the lot is 4 hours, meaning that $\mu = 0.25$. There are 10 parking spaces or servers and we do not allow cars to wait for an empty spot. Table 21.1 summarizes the key inputs.

Next, we create a table with columns corresponding to (a) the state number, n; (b) the state-dependent rate at which we transition up from state n to state $n + 1$, λ_n; (c) the state-dependent rate at which we transition down from state n to state $n - 1$, μ_n; (d) the non-normalized probability, which is nothing more than the state probabilities in terms of P_0, which

Table 21.1: Key inputs to the example finite queueing problem

Lambda	2
Mu	0.25
Max in system	10
Servers	10

Table 21.2: Numerical analysis of the example finite queueing problem

	A	B	C	D	E
7			Total NNP	2432.12127	
8					
9	State	Rate Up	Rate Down	Non Normalized Probability	Probability
10	0	2	0.00	1.00	0.00041
11	1	2	0.25	8.00	0.00329
12	2	2	0.50	32.00	0.01316
13	3	2	0.75	85.33	0.03509
14	4	2	1.00	170.67	0.07017
15	5	2	1.25	273.07	0.11228
16	6	2	1.50	364.09	0.14970
17	7	2	1.75	416.10	0.17109
18	8	2	2.00	416.10	0.17109
19	9	2	2.25	369.87	0.15208
20	10	0	2.5	295.89	0.12166

we will (arbitrarily) set to 1; and (e) the true state probabilities. Table 21.2 shows the results of this analysis while Table 21.3 shows the underlying Excel formulae used in this model.

At this point, a few observations are worth making. First, the upward transition rate from state 10, when the system is full, is 0. Second, as expected, the downward transition rate increases and is equal to $n\mu$ in all states. Third, we arbitrarily set $P_0 = 1$ in column D and then solve for all other probabilities using $P_n = \frac{\lambda_n-1}{\mu_n} P_{n-1}$, where $\lambda_n - 1$ is the state-dependent rate up from state $n - 1$ and μ_n is the state-dependent rate down from state n. The state-dependent upward transition rates are given in cells B10:B20 above in Table 21.2 while cells C10:C20 give the state-dependent downward rates. Next, in cell D7, we compute the sum of these non-normalized state probabilities. Finally, in column E, we compute the true probabilities by renormalizing the non-normalized probabilities of column D. The values in column E sum to 1.0.

Table 21.3: Equations corresponding to the numerical results of Table 21.2

	A	B	C	D	E
7			Total NNP	=SUM(Non_Normalized_Probability)	
8					
9	State	Rate Up	Rate Down	Non Normalized Probability	Probability
10	0	=IF(A10<Max_In_System,Lambda,0)	=MIN(A10,Servers)*Mu	1	=D10/Total_NNP
11	=A10+1	=IF(A11<Max_In_System,Lambda,0)	=MIN(A11,Servers)*Mu	=D10*B10/C11	=D11/Total_NNP
12	=A11+1	=IF(A12<Max_In_System,Lambda,0)	=MIN(A12,Servers)*Mu	=D11*B11/C12	=D12/Total_NNP
13	=A12+1	=IF(A13<Max_In_System,Lambda,0)	=MIN(A13,Servers)*Mu	=D12*B12/C13	=D13/Total_NNP
14	=A13+1	=IF(A14<Max_In_System,Lambda,0)	=MIN(A14,Servers)*Mu	=D13*B13/C14	=D14/Total_NNP
15	=A14+1	=IF(A15<Max_In_System,Lambda,0)	=MIN(A15,Servers)*Mu	=D14*B14/C15	=D15/Total_NNP
16	=A15+1	=IF(A16<Max_In_System,Lambda,0)	=MIN(A16,Servers)*Mu	=D15*B15/C16	=D16/Total_NNP
17	=A16+1	=IF(A17<Max_In_System,Lambda,0)	=MIN(A17,Servers)*Mu	=D16*B16/C17	=D17/Total_NNP
18	=A17+1	=IF(A18<Max_In_System,Lambda,0)	=MIN(A18,Servers)*Mu	=D17*B17/C18	=D18/Total_NNP
19	=A18+1	=IF(A19<Max_In_System,Lambda,0)	=MIN(A19,Servers)*Mu	=D18*B18/C19	=D19/Total_NNP
20	=A19+1	=IF(A20<Max_In_System,Lambda,0)	=MIN(A20,Servers)*Mu	=D19*B19/C20	=D20/Total_NNP
21	=A20+1	=IF(A21<Max_In_System,Lambda,0)	=MIN(A21,Servers)*Mu	=D20*B20/C21	=D21/Total_NNP

Table 21.4: Performance metrics for the example finite queue

L	7.03
Lambda Eff	1.76
W	4.00
Wq	0.00
Lq	0.00

Now we are ready to solve for the performance metrics of the system. The results of doing so are shown in Table 21.4 while the equations used to derive these results are shown in Table 21.5. The key difference between this analysis and the analyses of Chapters 18 and 20 is that we now need to compute the *effective arrival rate*. This is just given by $\lambda_{eff} = \sum_{all\,n} \lambda_n P_n$. It is worth noting that while the nominal arrival rate is $\lambda = 2$, some fraction of the customers (about 12% in this case) arrive when the system is full; they are effectively invisible to the system. Hence the effective arrival rate is only 1.76. It is this effective arrival rate that we need to use in relating the average number in the system, L, to the average time in the system, W and the average number waiting for service, L_q to the average time spent waiting for service, W_q. In other words, we need to use $L = \lambda_{eff} \cdot W$ and $L_q = \lambda_{eff} \cdot W_q$. Also note that the average time spent waiting is 0 as is the average number waiting. This is as expected since we do not allow vehicles to wait for empty parking spaces.

This numerical approach to solving the steady state balance equations can be extended to other more complex Markovian queues, including queues in which customers renege on the service, for example.

It is important to note that for queues with a finite state space, there is no requirement that the utilization ratio be less than 1. This should be clear from the non-normalized probabilities,

Table 21.5: Equations used for the performance metrics of the finite queue

L	=SUMPRODUCT(State,Probability)
Lambda Eff	=SUMPRODUCT(Rate_Up,Probability)
W	=L/Lambda_Eff
Wq	=W–(1/Mu)
Lq	=Wq*Lambda_Eff

which will always add up to a finite number if there are a finite number of states. Also, we can set the non-normalized value of P_0 to any value. Sometimes setting it to a value other than 1 will make the other non-normalized values nice integers.

21.3 SIMULATING THE BEHAVIOR OF A QUEUE

While the numerical method outlined in Section 21.2 works well for simple Markovian queues, or queues with exponentially distributed inter-arrival times and exponentially distributed service times, many systems exhibit more complex behavior. To analyze such systems, we often need to resort to simulation. A computer simulation is a program that attempts to replicate the important features of a real-world system. Simulation enables engineers and managers to experiment with policies, some of which may be very costly, without implementing them in the real world. The simulation models we discuss below are *stochastic* simulation models, in the sense that some element(s) of the model are described probabilistically. As such, a single run of the simulation model will give only one sampled estimate of the response. For example, as we will see below, we can simulate an M/M/1 queue. The result will give us one realization of the average time spent waiting. If we rerun the same model a second time, we will get a different estimate of the average waiting time, because the realized inter-arrival times and the service times are stochastic and will change from one run to the next.

Before going further with simulation modeling, we need to learn how to sample from a distribution. We will assume that the cumulative distribution function is known and that we can invert it. This is a very mild assumption and will apply to most distributions of interest. We sample from a continuous Uniform distribution to obtain a value between 0 and 1. The RAND() function in Excel is one way to do this. We then set this sampled value equal to the cumulative distribution and solve for the value of the random variable that results in this cumulative value for the cumulative distribution.

We illustrate this process with the exponential distribution, as shown in Figure 21.1. The blue line is the cumulative exponential distribution when the parameter of the distribution is 0.8. In other words, the equation of the blue line is $F_T(t) = 1 - e^{-0.8t}$. The figure shows us sampling twice (once with the red line and once with the green line). When we draw the first random

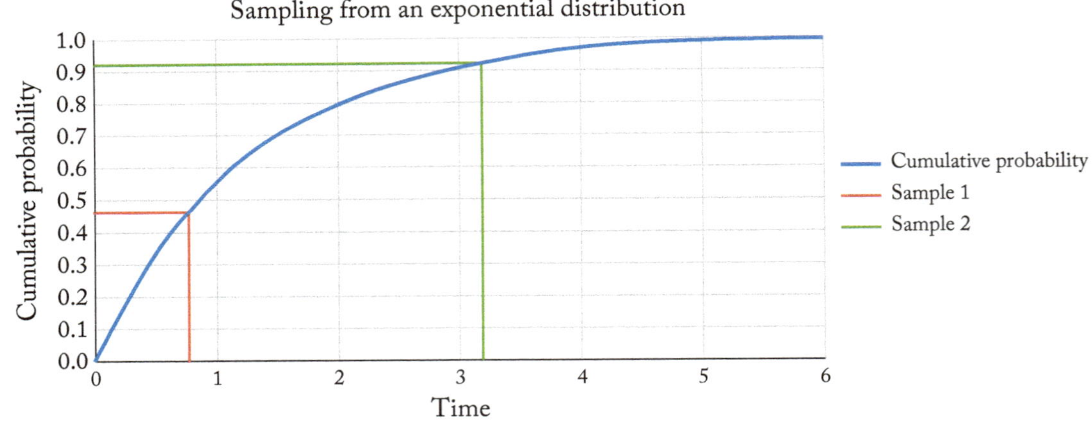

Figure 21.1: Sampling from an exponential distribution.

value equal to 0.461 (as shown in red), we set that value equal to the cumulative distribution, read across to the blue line and then down to the time axis to get a time of 0.773. When we do this a second time, we may get a value of 0.922 for the random point on the cumulative distribution. This corresponds to a time of 3.187 (as shown by the green line). Thus, the two samples from the exponential distribution in this case would be 0.773 and 3.187.

Analytically, we let u be a sample from the Uniform distribution. We then have $u = 1 - e^{-0.8t}$. When we solve for t, we obtain, $t = \frac{-\ln(1-u)}{0.8}$. Since u and $1 - u$ are both uniformly distributed between 0 and 1, we can simplify this to $t = \frac{-\ln(u)}{0.8}$.

This form avoids one arithmetic operation. If we are sampling a million times from a distribution using Excel, this change can save some time.

We are now ready to develop a simulation model. In Excel, we would set up six columns with labels:

1. Customer

2. Arrival time

3. Service time

4. Start service

5. End service

6. Wait time

Table 21.6 shows the first 10 customer arrivals for a single run of a simulation of an M/M/1 queue with $\lambda = 0.8$ and $\mu = 1$, while Table 21.7 shows the corresponding Excel statements. The

Table 21.6: Sample simulation of an M/M/1 queue

	Customer	Arrival Time	Service Time	Start Service	End Service	Wait Time	Running Total Wait Time	Running Avg Waiting Time
11	1	1.560501	0.05164843	1.560501	1.61214943	0	0	0
12	2	3.03354733	0.41416251	3.03354733	3.44770985	0	0	0
13	3	4.68483174	0.41999713	4.68483174	5.10482887	0	0	0
14	4	5.65942448	0.5425322	5.65942448	6.20195667	0	0	0
15	5	9.89083347	1.28447511	9.89083347	11.1753086	0	0	0
16	6	10.710515	0.49636016	11.1753086	11.6716687	0.46479362	0.46479362	0.0774656
17	7	11.62657	0.86605656	11.6716687	12.5377253	0.04509875	0.50989236	0.07284177
18	8	12.139737	3.16344934	12.5377253	15.7011747	0.39798829	0.90788065	0.11348508
19	9	12.1661791	1.68381932	15.7011747	17.384994	3.53499557	4.44287622	0.49365291
20	10	13.1021292	1.43115592	17.384994	18.8161499	4.28286475	8.72574097	0.8725741

Table 21.7: Sample code for a simulation of an M/M/1 queue

	Customer	Arrival Time	Service Time	Start Service	End Service	Wait Time
11	1	=-LN(1-RAND())/Lambda	=-LN(1-RAND())/Mu	=F11	=H11+G11	=H11-F11
12	=E11+1	=F11-LN(1-RAND())/Lambda	=-LN(1-RAND())/Mu	=MAX(F12,I11)	=H12+G12	=H12-F12
13	=E12+1	=F12-LN(1-RAND())/Lambda	=-LN(1-RAND())/Mu	=MAX(F13,I12)	=H13+G13	=H13-F13
14	=E13+1	=F13-LN(1-RAND())/Lambda	=-LN(1-RAND())/Mu	=MAX(F14,I13)	=H14+G14	=H14-F14
15	=E14+1	=F14-LN(1-RAND())/Lambda	=-LN(1-RAND())/Mu	=MAX(F15,I14)	=H15+G15	=H15-F15
16	=E15+1	=F15-LN(1-RAND())/Lambda	=-LN(1-RAND())/Mu	=MAX(F16,I15)	=H16+G16	=H16-F16
17	=E16+1	=F16-LN(1-RAND())/Lambda	=-LN(1-RAND())/Mu	=MAX(F17,I16)	=H17+G17	=H17-F17
18	=E17+1	=F17-LN(1-RAND())/Lambda	=-LN(1-RAND())/Mu	=MAX(F18,I17)	=H18+G18	=H18-F18
19	=E18+1	=F18-LN(1-RAND())/Lambda	=-LN(1-RAND())/Mu	=MAX(F19,I18)	=H19+G19	=H19-F19
20	=E19+1	=F19-LN(1-RAND())/Lambda	=-LN(1-RAND())/Mu	=MAX(F20,I19)	=H20+G20	=H20-F20

first row is often slightly different from the successive rows. Cell F11 computes the arrival time of the first customer, for example. Cell F12 computes the arrival time of the second customer as the arrival time of the first customer plus an exponentially distributed amount of time. Subsequent entries in column F are similar to cell F12. Column G computes the sampled service times. Cell H11 gives time the first customer start service. Since we assume the system is empty and idle at the beginning of the simulation, this customer begins service as soon as she arrives. She ends her service at a time equal to the start time of her service plus her service time as shown in cell I11. Subsequent customers do not necessarily start service as soon as they arrive. They start at the larger of their arrival time and the time the preceding customer finishes service as shown in the formulae in cells H12 and below. In fact, customer 6 arrives at time 10.71, but customer 5

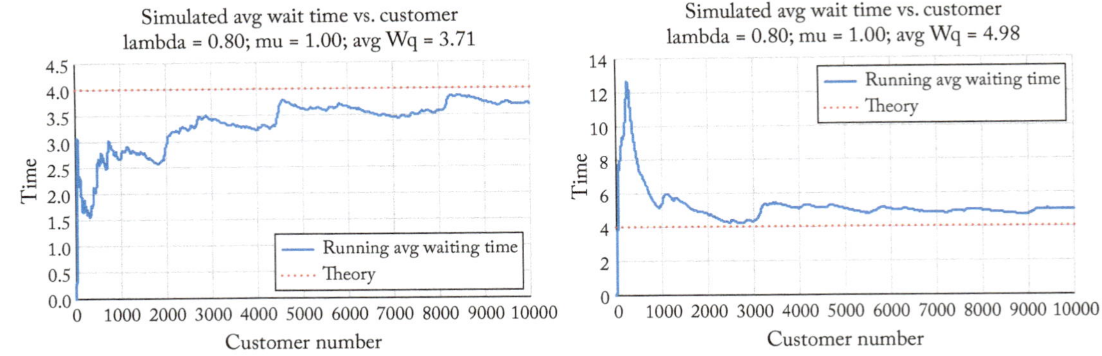

Figure 21.2: Two simulation runs for an M/M/1 queue with $\lambda = 0.8$ and $\mu = 1$.

does not finish service until time 11.175, and so customer 6 cannot start his service until that time. The wait time of any customer is the difference between when the customer starts service and when the customer arrives. Customer 6 waits about 0.465 time units, for example, as shown in Table 21.6.

These tables only show the first 10 customer arrivals. In fact, the results discussed below are based upon a simulation of 10,000 customer arrivals. The simulation runs through row 10,010 of the Excel spreadsheet.

Figure 21.2 plots two different runs of the simulation model with $\lambda = 0.8$ and $\mu = 1$. The blue line represents the running average waiting time. This is just the average waiting time that the model would report if the simulation were to be terminated at the given number of customers. For example, in the left panel, if the simulation ended after 2,000 customer arrivals, the reported average time would be about 2.92. In the right panel, if the simulation model were to be terminated after 2,000 arrivals, the model would return an average waiting time of about 4.78. The red line at 4.0 in each graph is the theoretical value. In the simulation run shown on the left, the average time never gets up to the theoretical value. In the run on the right, the average never gets below the theoretical value after the first few arrivals. The simulation run on the left would predict a waiting time of 3.71 after 10,000 customer arrivals, while the run on the right would predict a waiting time of 4.98. **This clearly indicates that the output of a simulation model is a random variable.**

Finally, let us consider the results shown in Figure 21.3. In all cases, we used a mean service time of 1.0 or $\mu = 1$. We varied the arrival rate to obtain the 9 utilization ratios of 0.55, 0.60, ..., 0.95. For each utilization level we ran the simulation model 10 times, simulating 10,000 customer arrivals each time. In other words, we simulated a total of 900,000 customer arrivals. The entire process takes seconds on a desktop computer running Excel. The results are shown as 10 blue dots for each utilization level.

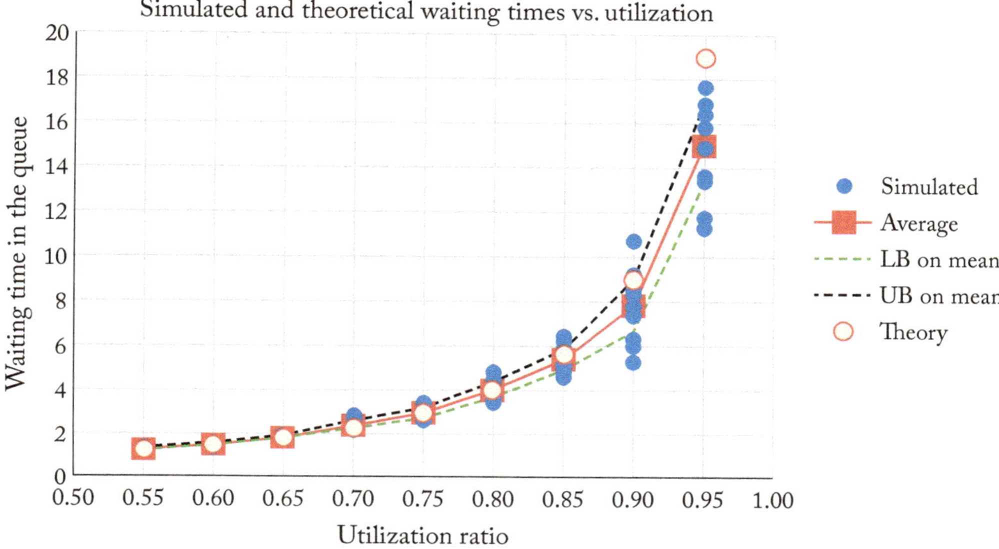

Figure 21.3: Ten simulated waiting times for nine utilization levels.

The first observation is that the variability of the simulated average waiting times increases with the utilization level. The range of the 10 simulated values with $\rho = 0.55$ is 0.17, while the range for $\rho = 0.95$ is 7.62. This is consistent with our result in Chapter 18 in which we showed that the standard deviation of the number of customers in the system was inversely proportional to $1 - \rho$.

In addition, the graph plots the average of the 10 simulation runs with a red square, the estimated 95% confidence interval for the mean with the dashed lines, and the theoretical value for each of the nine values of the utilization level. In 7 of the 9 cases, the 95% confidence interval (which is actually a random interval) includes the theoretical mean shown with an orange circle. When $\rho = 0.95$ the 95% confidence interval goes from 13.30 to 16.89, but the theoretical waiting time is 19.0. The 10 simulated values tend to consistently underestimate the true mean in this case. When $\rho = 0.55$, the 95% confidence interval extends from 1.223 to 1.295, but the theoretical time is only 1.222. In this case, the simulation tends to overestimate the true waiting time. Such results are not unexpected as the simulation model outputs are random variables. Incidentally, the reader should not infer anything from the fact that in this case, the simulation model underestimated the waiting time at the highest utilization level shown and overestimated the waiting time at the lowest utilization level shown. This is purely by chance.

21.4   SUMMARY

In this chapter, we have outlined two numerical methods for solving for the performance of queueing systems. The non-normalized probability method shown in Section 21.2 can readily be extended to more complicated cases as long as the system remains Markovian, or, for our purposes, as long as the inter-arrival times are exponentially distributed and the service completion times are also exponentially distributed. For example, we can use this method to model multiple server systems, with finite capacity, state-dependent arrival rates, and reneging, if the distributions are all exponential.

When the relevant distributions are not exponential, we can use simulation. This is a very versatile tool and much has been written about it. The interested reader should see such texts as Banks et al. [2010] and Nelson [1995]. In all cases, however, it is important to remember that the output of a simulation model is itself a random variable and is subject to variability. Multiple long runs may be needed to get accurate predictions of key output metrics, particularly for high utilization levels of the system resources.

21.5   REFERENCES

J. Banks, J. S. Carson II, B. L. Nelson, and D. M. Nicol, *Discrete-Event System Simulation*, 5th ed., Prentice Hall, Upper Saddle River, NJ, 2010. 152

M. S. Daskin, *Service Science*, John Wiley & Sons, New York, 2010. DOI: 10.1002/9780470877876. 144

B. L. Nelson, *Stochastic Modeling: Analysis and Simulation*, Dover Publications, Mineola, NY, 1995. 152

C H A P T E R   22

Decision Theory

22.1   UNCERTAINTY vs. RISK

We have suggested throughout this book that operations management is about making *decisions* in the face of uncertainty to improve or *optimize* the performance of a system, while accounting for important *tradeoffs*. In doing so, we have used uncertainty in a rather casual manner allowing it to refer to probabilistic or uncertain situations. For example, the demand for a product, or the number of snowfalls during a season, or the number of customers in a service system at any point in time, were all considered uncertain, even if they could be described by a probability distribution. For example, we showed that the number of customers in an M/M/1 queue followed a geometric distribution and we said that there was uncertainty surrounding the number in the queue at any point in time.

Decision theory is an important branch of operations management. Decision theory helps us structure the process of making decisions in the face of unpredictable future conditions. Decision theory distinguishes between *uncertainty* and *risk* [Eiselt and Sandblom, 2012]. *Uncertainty* is used when there is no knowledge of the likelihood of future events. In other words, decision theorists say that the future is uncertain if they cannot associate probabilities with future conditions. *Risk*, on the other hand, examines situations in which a probabilistic description of future events can be stated. For example, if all we can say is that the demand for a product will be between 0 and 10 items, we would say the future is uncertain. If, on the other hand, we can say that there is a specific probability mass function associated with each of the 11 demand levels, we would say that there is risk associated with the demand.

In this chapter, we will introduce basic concepts of decision theory. We will do so in the context of the newsvendor problem, which should be familiar to you by now. We will begin by casting the traditional newsvendor problem as a decision theoretic problem with risk. Then we will introduce the concept of *regret* and show how that can be used in the context of the newsvendor problem. Finally, we will show that these two concepts lead to a *tradeoff* between maximizing the expected profit and minimizing the maximum regret.

Table 22.1: Probability mass function of demand for a camera

Demand	Probability Mass Function of Demand	Cumulative Distribution of Demand
0	0.01	0.01
1	0.04	0.05
2	0.12	0.17
3	0.17	0.34
4	0.19	0.53
5	0.21	0.74
6	0.15	0.89
7	0.05	0.94
8	0.03	0.97
9	0.02	0.99
10	0.01	1.00

22.2   THE NEWSVENDOR PROBLEM IN DECISION THEORY TERMS

In decision theory, we must make a *decision* in the face of uncertain or risky conditions. Following that decision, (a part of) the future uncertainty is revealed as the outcome of a *random event*. This, in turn, leads to *outcomes*.

In the context of the newsvendor problem, we must decide now how many items to buy. Following that the demand is realized. For every combination of the number of items bought and the demand, there is an outcome, which is the profit as a result of the decision regarding the number to buy and the demand.

To illustrate this, suppose we own a small camera store. Yes, such stores are a vanishing breed, as mail-order operations and Amazon are rapidly eating into their profits, but this will give us a lens through which we can expose new concepts, zoom in on different ideas, and develop the landscape of decision theory. I shutter to think what comes next.

We need to place an order for a new moderately priced camera, which we know, will rapidly become outdated. Table 22.1 gives the probability mass function of the demand for the camera. We can purchase the camera for $500 from the manufacturer; we sell it for $1,200; and unsold cameras when the new model comes out can be returned to the manufacturer for a refund of $250. The critical ratio is

$$\frac{1200 - 500}{1200 - 250} = \frac{700}{950} \approx 0.737$$

so the optimal number to buy is 5. This will result in an optimal expected profit of $2,455.00.

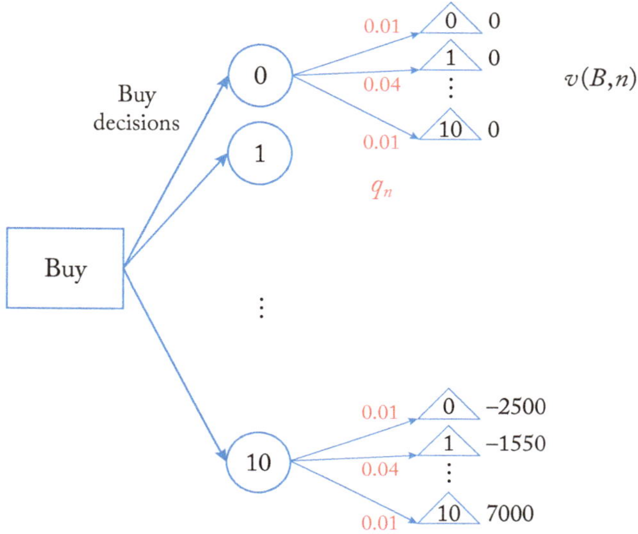

Figure 22.1: Decision tree for example newsvendor problem.

Figure 22.1 shows how this problem can be cast as a decision theoretic problem. Rectangles are used to represent decisions, circles represent random events, and triangles represent outcomes. In this example, there is only one decision, the number of cameras to buy. This is followed by one of 11 random events representing the actual demand. The value or reward associated with buying B items and having a demand of n items is given by $v(B,n) = -cB + p\min(B,n) + s\max(0, B - n)$, where c is the unit cost of buying an item from the manufacturer or supplier ($500 in this case), p is the unit sale price (or $1,200 in this case), and s is the unit salvage value (or $250 in this case). The values of $v(B,n)$ are shown in black on the right-hand side of the decision tree in Figure 22.1. We also associate a probability, q_n, with each level of demand. These values are shown in red beside the links from a random event node to an outcome node.

By taking the expected value of the value or reward for a given number of items to buy, $v(B) = \sum_{n=0}^{10} q_n v(B,n)$, we can obtain the value associated with buying B items. Figure 22.2 shows the resulting collapsed decision tree. The figure also explicitly shows the optimal number of items to buy, which is 5 in this case. The expected profit is $2,455.00. Figure 22.3 plots the expected profit vs. the number of items to buy and highlights that the optimal number to buy is 5.

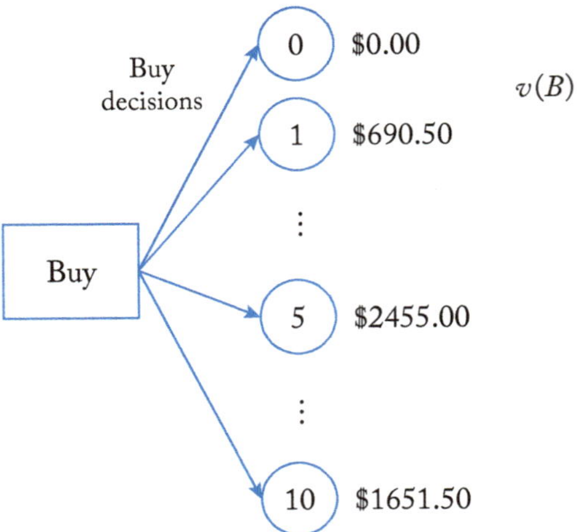

Figure 22.2: Collapsed decision tree showing the expected profit associated with each decision.

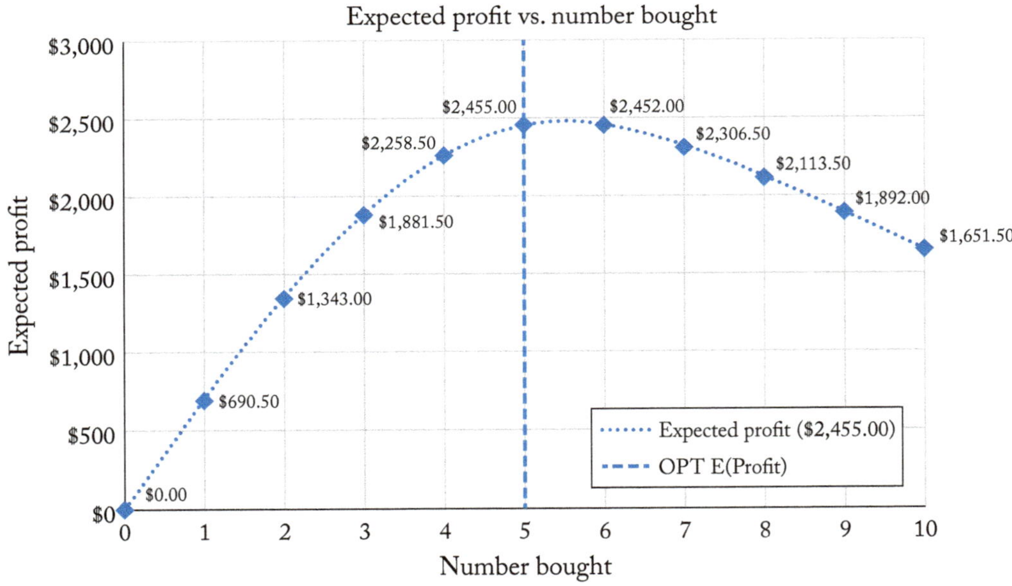

Figure 22.3: Expected profit vs. number of items to buy.

22.3   REGRET

The approach outlined in Section 22.2 is typical of the risk-based approach to decision theory. We are maximizing the *expected value* associated with the decision regarding how many cameras to buy. When we can accurately estimate the probability mass function of the demand, this may be an appropriate approach. When we cannot do so, other approaches must be adopted.

Whether or not we can predict the probability mass function, however, we may also want to optimize the *regret* associated with the decision we make. The regret is a measure of how bad we will feel about a decision after the actual demand is realized. If we buy 5 cameras, for example, and the demand is only 3, we will say to ourselves, "If I had only known that the demand was going to be 3, I would have bought 3 cameras and my profit would be \$2,100 (or \$3,600 − \$1,500) instead of the \$1,600 that I am now getting (\$3,600 − \$2,500 + \$500)." Your regret in this case would be \$500 or \$2,100 − \$1,600. Similarly, if we buy 5 cameras and the demand turns out to be 8, our regret would be \$2,100. We would say to ourselves, "If I had only known the demand was going to be 8, I would have bought 3 more cameras. I would have made an extra \$700 on each camera or a total of \$2,100."

Specifically, we can define the regret associated with the decision to buy B cameras when the demand is n as

$$R(B,n) = -cn + pn - \{-cB + p\min(B,n) + s\max(0, B - n)\}.$$

The first two terms, $-cn + pn$, represent the profit we would have made had we known that the demand was going to be n. The term in braces is the profit we make when we buy B items and the demand is n.

It turns out that minimizing the expected regret is equivalent to maximizing the expected profit. However, we may want to *minimize the maximum* (or worst-case) *regret*. After all, if the regret is very small, we will not be too concerned about it. If the regret is very large, we will look really bad and we will seriously wish we had better predicted the demand. This is often the case. For example, in locating ambulances, we are likely to be more concerned with the worst-case response time than we are with the average response time. After all, it is the people who experience exceptionally long response times who are most likely to complain, not those who experience very short response times. Similarly, in assigning students to seminars, it is likely to be those students who get their fifth or sixth choice seminar who will complain, and not those who get their first or second choice seminar.

For any number of items we buy, the maximum regret will occur when either the demand is 0 or the demand is the maximum possible demand, which we denote by $D_{\max}$. Specifically, we will have

$$R(B,0) = cB - sB = (c - s)B$$

and

$$R(B, D_{\max}) = c(B - D_{\max}) + pD_{\max} - pB.$$

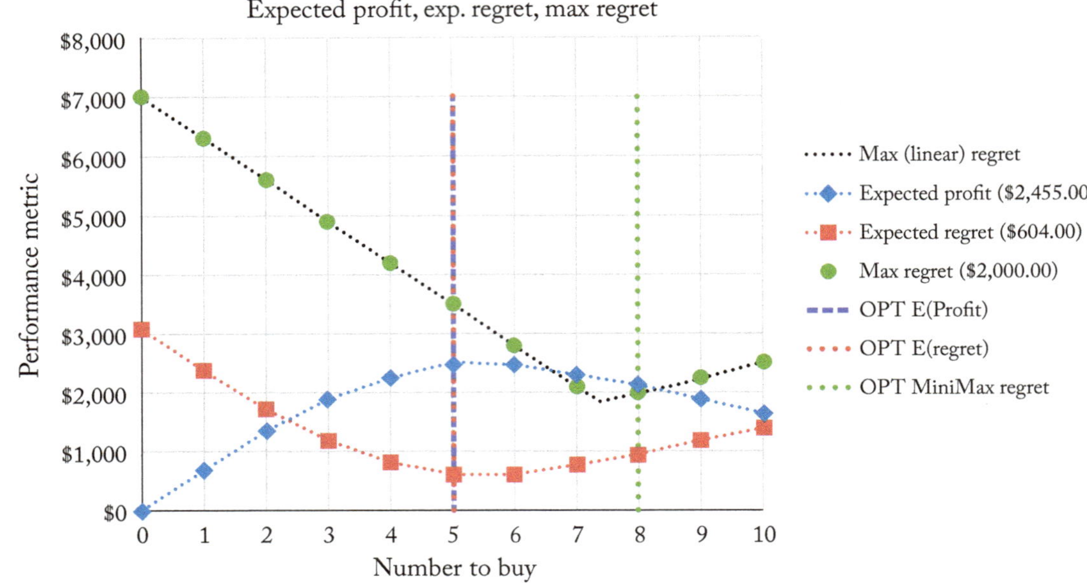

Figure 22.4: Expected profit, expected regret, and maximum regret vs. number to buy.

For large purchase sizes ($B \geq \frac{p-c}{p-s} D_{\max}$), the first equation will apply; for small purchase sizes ($B \leq \frac{p-c}{p-s} D_{\max}$), the second equation will apply.

Figure 22.4 illustrates these concepts. The horizontal axis shows the number of items to buy and the vertical axis is the performance metric. The blue diamonds connected by a dashed line represent the expected profit. This curve is identical to the curve shown in Figure 22.3. The red squares represent the expected regret. Note that this curve attains its minimum at exactly the same point that the expected profit curve attains its maximum. The black dashed V-shaped line shows the maximum regret as a function of the number of items to buy. The minimum occurs at $B = 7.37$. Naturally, we cannot buy a fractional number of cameras and so we look instead at the regret given by the discrete green circles. Specifically, we look at the regret associated with $B = 7$ which is $2,100 and the regret associated with $B = 8$ which is $2,000 and select the smaller of these two values. Thus, if we want to minimize the maximum regret, we should buy eight cameras (shown with the vertical green dashed line), while maximizing the expected profit or minimizing the expected regret results in our buying five cameras (shown with the vertical red or blue dashed lines).

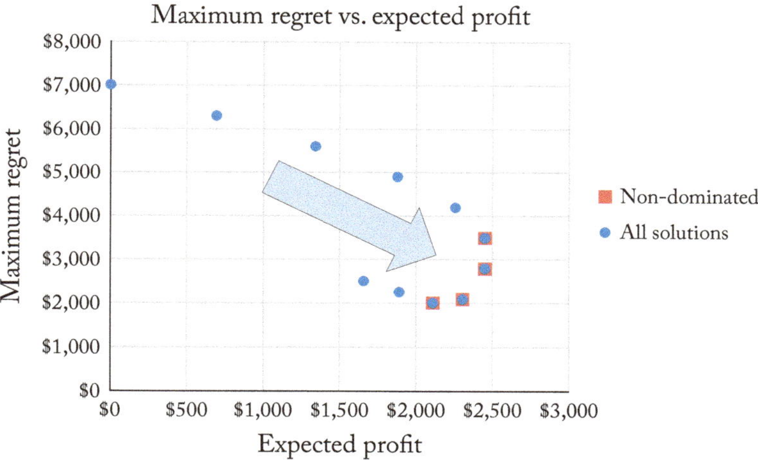

Figure 22.5: Maximum regret vs. expected profit.

22.4   COMING FULL CIRCLE: ANOTHER TRADEOFF

In Section 22.2, we maximized the expected profit. In Section 22.3, we introduced the concept of regret and showed that minimizing the expected regret was equivalent to maximizing the expected profit. We showed that minimizing the worst-case or maximum regret resulted in a different decision from the one suggested by maximizing the expected profit. Thus, we have addressed *optimizing* the decision in the face of *uncertainty* or *risk*. We have addressed two of the three fundamental issues addressed by operations management. In this section, we address the third issue, that of a *tradeoff*.

In Figure 22.5, we plot the maximum regret vs. the expected profit for every one of the 11 possible buy decisions. These are shown as small blue circles in the figure. Since we want to maximize the expected profit and minimize the maximum regret, the arrow points in the direction of the ideal solution. The four solutions, highlighted as orange squares, are non-dominated solutions. Any of the other seven solutions are inferior to one or more of these four solutions. Figure 22.6 highlights these four solutions. The top-right solution is the solution that maximizes the expected profit, or the solution found in Section 22.2. The left-most solution minimizes the maximum regret. This is the solution found in Section 22.3. The other two solutions, however, represent good compromise solutions. For example, the third solution from the left (with coordinates $2,452 and $2,800) decreases the expected profit by $3.00 or 0.12% while reducing the maximum regret by $700 or 20% of the value associated with the solution found in Section 22.2. This solution corresponds to buying six cameras instead of five. Giving up $3.00 on average to reduce the worst-case regret by $700 may be a very good decision.

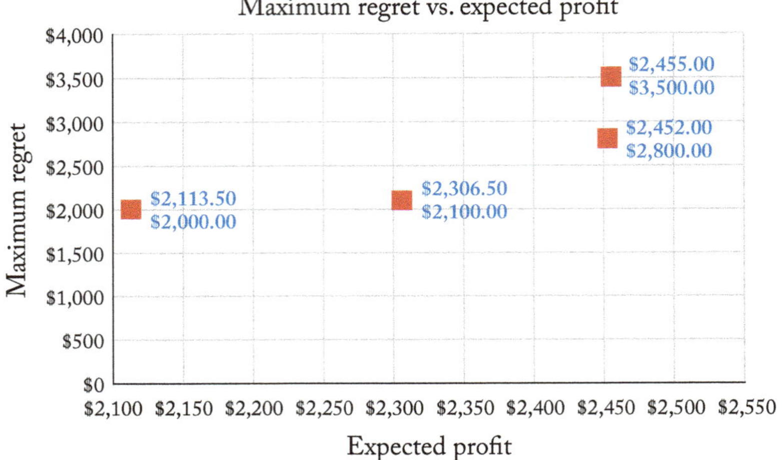

Figure 22.6: The tradeoff between maximum regret and expected profit.

As before, it is important to consider the full range of solutions represented by the tradeoff between different objectives before selecting a single decision to implement.

22.5 REFERENCES

H. A. Eiselt and C.-L. Sandblom, *Operations Research: A Model–Based Approach*, 2nd ed., Springer, Heidelberg, Germany, 2012. DOI: 10.1007/978-3-642-31054-6. 153

CHAPTER   23

Summary, Where to Next, and Final Thoughts

23.1   YOU HAVE COME A LONG WAY

At the beginning of this text, we argued that operations management is about finding improved or *optimal* decisions in the face of *uncertainty* while accounting for important *tradeoffs*. Throughout the text, we have tried to emphasize these key themes. We have introduced key methodologies and a number of important classes of problems that can be addressed using operations management. The goal has been to **introduce** you to key topics and hopefully to entice you to learn more about operations management.

For those with little or no exposure to probability theory, Chapters 2–6 introduced key concepts of probability. We began by distinguishing between probability and its sister methodology, statistics. We looked at conditional probability and Bayes' Theorem. In Chapter 3, we tried to develop some intuition about probability and suggested that our natural intuition was often misleading. Chapters 4 and 5 introduced various discrete and continuous distributions that are often used in probabilistic modeling. Chapter 6 summarized key functions of random variables including the mean, variance, and standard deviation of a random variable. A key point in Chapter 6 is that unless the function is linear, the expected value of the function is generally *not* equal to the function evaluated at the expected value of the random variable.

We then turned to optimization. Chapter 7 introduced the basics of optimization: inputs, decision variables, objective(s), and constraints. We argued that sorting out these four issues was at least half the battle in any optimization modeling exercise. We applied these concepts to the problem of linear regression, or fitting a line to a set of observations. Chapter 8 applied these concepts to the most fundamental of all inventory problems, the economic order quantity model. This model finds the optimal tradeoff between the cost of placing orders which goes down as the order quantity goes up and the holding cost of goods which goes up with the order quantity. In Chapter 9, we applied basic concepts of uncertainty analysis to the problem of finding the optimal number of items to buy when the demand for the product is uncertain. Here we were trading off the cost associated with having to salvage items that we cannot sell against the loss of revenue that comes with not having enough inventory to satisfy demand.

Chapters 10, 11, and 12 dealt with linear programming, a key methodology in operations management. Chapter 10 introduced linear programming graphically. This allowed us to visualize the feasible region of a linear programming problem as well as contours of the objective function. We showed that at least one optimal solution occurs at a corner of the feasible region. We also showed that the newsvendor problem of Chapter 9 could be cast as a linear (network flow) programming problem. Chapters 11 and 12 dealt with two applications of linear programming: relocating rental vehicles to minimize the relocation cost and assigning students to seminars to maximize student preferences.

Chapters 13–16 dealt with location modeling, another key application domain of operations management. Chapter 13 introduced an analytic model, which is similar to, though different from, the economic order quantity model. In this model, we find the optimal tradeoff between the cost of the facilities which increases with the number of facilities and the transportation cost to serve the customers which decreases with the number of facilities. Chapter 14 introduced the concept of coverage in discrete location models and Chapter 15 discussed the demand-weighted average distance. Both problems were cast as integer linear programming problems. Finally, in Chapter 16 we discussed the tradeoff between these two objectives.

Chapters 17–21 discussed queueing theory or the mathematics behind the formation of lines for service. Key assumptions and inputs were outlined in Chapter 17. The chapter also introduced four key performance metrics for queues and the relationships between those measures. Chapter 18 discussed the simplest of all stochastic queueing models in which we have Poisson arrivals (or exponentially distributed inter-arrival times), exponentially distributed service times, and a single server. We showed that the performance of a queueing system degrades rapidly as the utilization level increases beyond about 0.8. In Chapter 19, we introduced a model that allows the service time distribution to be (just about) any distribution we like. This allowed us to show that performance degrades as the variability of the service times increases. In Chapter 20, we reverted to exponentially distributed service times, but allowed the number of servers to be any number we wanted. Finally, Chapter 21 introduced a numerical method for solving the steady state balance equations used in Chapters 18 and 20 and introduced simulation as a more general way of handling queueing problems. We showed that the variability of a simulation model outputs increases as the utilization of the system increases.

Chapter 22 introduced decision theory and distinguished between models of risk and those associated with pure uncertainty. We introduced the notion of regret. Again, we used the newsvendor problem as an example and showed that there was an inherent tradeoff between maximizing the profit and minimizing the maximum regret associated with our decisions.

You have indeed come a long way!

23.2   MUCH MORE TO LEARN

That said, there is much more to learn. Perhaps the most important concept that was not discussed in this text is that of *duality* in linear programming. Underlying every linear programming problem is an associated dual linear programming problem. The values of the dual variables give us the "value" associated with having more of any given resource. Clearly, if we are not using all of a resource, the value of having more should be 0, meaning that the value of the dual variable associated with that resource constraint should be 0. That said, there is **much** more to learn about duality in linear programming.

In addition, we have not touched on any of the methods used to solve linear programming problems, network flow problems, or integer linear programming problems. We have not discussed many other example problems, including the shortest path problem, which underlies all route guidance algorithms. We have not talked about dynamic programming, nor have we said much about the important areas of stochastic programming or robust optimization.

We have just scratched the surface of decision theory and we have said nothing about the associated field of game theory, which outlines how two or more agents interact.

The list of omitted topics could go on and on. Hopefully, the text has inspired you to learn more about these topics. Any of the general operations management, operations research, or management science texts listed in the references would be good starting points for this sort of exploration. In addition, there are many specialized texts dealing with particular topics and modeling approaches.

There really is a lot further we can go!

23.3   ALL MODELS ARE WRONG...

At several points in the text we have mentioned the famous quote attributed to George Box that says that "All models are wrong; but some are useful." Before closing, it is important to reflect on this in the context of a book on operations management. First it suggests that there is likely to be some legitimate criticism of just about any mathematical model we develop. For example, we may be approximating the service time distribution in a queueing model by the exponential distribution because this leads to mathematically tractable results even though the true distribution is somewhat different from the exponential distribution. In assigning students to classes or seminars, someone might suggest that assigning five students to their fourth choice as opposed to their third choice is much worse than assigning 105 students to their second choice instead of assigning only 100 students to their second choice. Note that the model of Chapter 12 would have both cases contributing 5 to the objective function. In location modeling, there are likely to be important cases in which individuals may not go to the closest facility, as assumed in most of the models we outlined.

Second, it should remind us that underlying *any* decision context is a set of models. When government officials decide whether or not to authorize a vaccine for emergency use, they are using some models. When other government officials decide that vaccinated individuals can stop wearing masks indoors, they too are using some models. When a university decides to admit 1,200 new students for the coming year, they too are using some models. In many of these cases, however, the models are not explicitly stated; rather, they are simply mental models. The value of the more rigorous mathematical models introduced in this text, is that they force the decision maker (1) to confront his or her assumptions, (2) to make those assumptions explicit, and (3) to be clear about what their objectives are. In doing so, they are providing others with a means of testing the implications of alternative assumptions and of employing different objectives. They allow for an explicit analysis of the tradeoffs between competing objectives. The value of this sort of rigor and or being explicit about the assumptions we are making and our objectives cannot be overestimated. Without this sort of rigor and without making our assumptions and objectives explicit, two or more people can believe that they are sharing views on a problem, when in fact they may be talking past each other. They may well be approaching the problem from very different perspectives. Even if we cannot "solve" the models we develop, the process of explicitly modeling the underlying problem is likely to be of significant value in and of itself as it facilitates useful communication.

23.4   REFERENCES

H. A. Eiselt and C.-L. Sandblom, *Operations Research: A Model-Based Approach*, 2nd ed., Springer, Heidelberg, Germany, 2012. DOI: 10.1007/978-3-642-31054-6.

J. Heizer and B. Render, *Operations Management*, Pearson, Boston, MA, 2011.

F. S. Hillier and G. J. Lieberman, *Introduction to Operations Research*, 6th ed., McGraw Hill, Boston, MA, 2010.

F. S. Hillier and M. S. Hillier, *Introduction to Management Science: A Modeling and Case Studies Approach with Spreadsheets*, Boston, MA, 2003.

P. A. Jensen and J. F. Bard, *Operations Research: Models and Methods*, John Wiley & Sons, Hoboken, NJ, 2003.

S. Nahmias, *Production and Operations Analysis*, 4th ed., McGraw Hill, Boston, MA, 2001.

R. S. Russell and B. W. Taylor III, *Operations Management: Focusing on Quality and Competitiveness*, Prentice Hall, Upper Saddle River, NJ, 1998.

M. K. Starr, *Operations Management: A Systems Approach*, Boyd and Fraser Pub. Co., Danvers, MA, 1996.

B. W. Taylor III, *Introduction to Management Science*, 10th ed., Prentice Hall, Upper Saddle River, NJ, 2010.

H. A. Taha, *Operations Research: An Introduction*, Pearson, Harlow, England, 2017.

H. Wagner, *Principles of Operations Research*, 2nd ed., Prentice Hall, Englewood Cliffs, NJ, 1975.

W. L. Winston, *Operations Research: Applications and Algorithms*, 4th ed., Brooks/Cole–Thomson Learning, Belmont, CA, 2004.

Index

Author's Biography

MARK S. DASKIN

Mark S. Daskin is the immediate past Department Chair of the Industrial and Operations Engineering Department at the University of Michigan. He holds the Clyde W. Johnson Collegiate Professorship. Prior to joining the faculty at Michigan in 2010, Daskin was on the faculty at Northwestern University (for 30 years) and the University of Texas (for a year and a half).

He received his Ph.D. from the Civil Engineering Department at M.I.T. in 1978. He also holds a B.S.C.E. degree from that department and a Certificate of Post-Graduate Study in Engineering from the University of Cambridge in England.

His research focuses on the application and development of operations research techniques for the analysis of health care problems, as well as transportation, supply chain, and manufacturing problems. He is the author of over 80 refereed papers and of two books: *Network and Discrete Location: Models, Algorithms and Applications* (John Wiley, 1995; second edition, 2013) and *Service Science* (John Wiley, 2010), winner of the IIE Joint Publishers Book of the Year Award in 2011.

Daskin received the Frank and Lillian Gilbreth Award from the Institute of Industrial and Systems Engineers in 2021. He was elected to the U.S. National Academy of Engineering in 2017. He is a Fellow of both INFORMS and IIE and has received the David F. Baker Distinguished Research Award, the Technical Innovation Award and the Fred C. Crane Award for Distinguished Service from IIE, as well as the Kimball Medal for service to the society and the profession from INFORMS. He received the Lifetime Achievement Award in Location Analysis from the Section on Location Analysis of INFORMS in 2014. He is a past editor-in-chief of both *IIE Transaction* and *Transportation Science*. In 2006, he was the president of INFORMS. He served as the chair of the Department of Industrial Engineering and Management Sciences at Northwestern University from 1995–2001.

CPSIA information can be obtained
at www.ICGtesting.com
Printed in the USA
BVHW010500141121
621428BV00001B/9